Healing with Artificial Intelligence

Diagnosis through images, robot surgeons, digital twins, and the metaverse are some of the applications in which artificial intelligence (AI) is involved. It is an almost sci-fi world that touches us closely and toward which we can harbor both great hopes and great fears. Health, in fact, is a central theme, and understanding how this new technology can help us to heal, acquire well-being, and live better is certainly important. This book, thanks to the author's great experience, deals comprehensively and accessibly with the benefits and risks of using AI in the field of health. It will enable the reader to confidently approach a revolution that will change the way we treat ourselves.

Daniele Caligiore is a Research Director at the Institute of Cognitive Sciences and Technologies, National Research Council of Italy (ISTC-CNR), where he works on AI for the study of the brain. He is the director and co-founder of the Advanced School in AI for the interdisciplinary study of artificial intelligence. He is also among the founding partners of AI2Life, a spin-off company of ISTC-CNR focused on AI solutions for social development and human well-being. Additionally, he teaches 'Artificial Intelligence Systems in Social Contexts' at the Libera Università Maria Santissima Assunta (LUMSA) in Rome. He has published more than 90 scientific articles in international journals, conference proceedings, and book chapters. He is the author of the interdisciplinary books *IA istruzioni per l'uso* (2022) and *Curarsi con l'Intelligenza Artificiale* (2024), both published by Il Mulino, as well as *Simulating the Brain: A Four-Step Method Using Ordinary Differential Equations and Python* (2025), published by Springer. He serves as an evaluator for the Italian Ministry of University and Research and the European Commission, reviewing research project proposals in the fields of artificial intelligence and neuroscience.

W0234815

Healing with Artificial Intelligence

Daniele Caligiore

CRC Press
Taylor & Francis Group
Boca Raton London New York

CRC Press is an imprint of the
Taylor & Francis Group, an **informa** business

Designed cover image: Getty Images

First edition published 2025
by CRC Press
2385 NW Executive Center Drive, Suite 320, Boca Raton FL 33431

and by CRC Press
4 Park Square, Milton Park, Abingdon, Oxon, OX14 4RN

CRC Press is an imprint of Taylor & Francis Group, LLC

© 2026 Daniele Caligiore

ISBN: 9781032998169 (hbk)
ISBN: 9781032991467 (pbk)
ISBN: 9781003606130 (ebk)

DOI: 10.1201/9781003606130

Typeset in Minion
by KnowledgeWorks Global Ltd.

to Serena and Chiara

Contents

Preface, ix

Acknowledgements, xiii

CHAPTER 1 ▪ What Artificial Intelligence Is and Its Historical
Applications in Medicine   1

  1.1 ARTIFICIAL INTELLIGENCE, MACHINE LEARNING,
AND DEEP LEARNING   1

  1.2 THE IMPORTANCE OF DIGITISING DATA   9

  1.3 MEDICAL IMAGING, ROBOTIC SURGERY, AND
TELEMEDICINE   11

  1.4 EXOSKELETONS AND CYBORGS   13

  NOTES   16

  REFERENCES TO CHAPTER ONE   17

CHAPTER 2 ▪ How AI Will Revolutionize Disease Diagnosis,
Treatment, and Prevention   19

  2.1 DATA-DRIVEN AI AND THEORY-DRIVEN AI   19

  2.2 DIGITAL TWINS, METAVERSE, AND ORGANOIDS   23

  2.3 AI FOR EARLY DIAGNOSIS   33

  2.4 AI FOR COMPUTER-BASED TESTING OF NEW
PERSONALIZED THERAPIES   38

  2.5 AI FOR DISEASE PREVENTION   42

  NOTES   44

  REFERENCES TO CHAPTER TWO   46

CHAPTER 3 ■ Ethics, Privacy, and Other Issues in AI
Medicine   51

3.1   WILL AI REPLACE DOCTORS?   51

3.2   WILL WE LOSE CONTROL OVER OURSELVES?   58

3.3   IS RESPONSIBILITY WITH AI OR THE PHYSICIAN?   63

3.4   DATA: PRIVACY, SECURITY, AND BIAS   66

3.5   UNDERSTANDING HOW THE AI MAKES CHOICES
      (EXPLAINABILITY)   68

NOTES   72

REFERENCES TO CHAPTER THREE   73

CHAPTER 4 ■ The Medicine of the Future: A Contamination
of Skills and Disciplines   75

4.1   TREATING THE DISEASE INDIVIDUALLY BUT
      LOOKING AT THE 'SYSTEM'   75

4.2   A NEW RENAISSANCE IN CONTAMINATION OF
      SKILLS AND DISCIPLINES   80

4.3   STUDYING A DISEASE WITHOUT GUINEA PIGS   87

4.4   TRAINING DOCTORS AND PATIENTS FOR MINDFUL
      USE OF AI   89

4.5   A HIGH QUALITY OF CARE DESPITE THE GROWING
      POPULATION   90

NOTES   93

REFERENCES TO CHAPTER FOUR   94

CONCLUSIONS, 95

INDEX, 99

Preface

In recent years, artificial intelligence (AI) has become increasingly pervasive in many aspects of our lives, from work and social interactions to everyday routines. We engage with AI daily, often without even realizing it. For instance, AI-powered natural language processing allows us to use voice commands on our smartphones to search for news on Google. Platforms like Spotify, Netflix, and Amazon utilize advanced AI algorithms known as recommendation systems to analyze our preferences and suggest music, films, or products tailored to our tastes. Many banks use similar algorithms to prevent financial fraud by analyzing our spending patterns. When unusual transactions occur, such as in unfamiliar locations or for atypical amounts, these systems detect potential risks and immediately notify the bank, which can then block credit cards to prevent fraud. Some online news platforms also leverage AI to generate simple articles, like sports reports or financial updates, enabling faster and more efficient dissemination of information.

The widespread adoption of AI is strongly supported by its integration with neurotechnology,[1] as well as information and digital technologies[2] [Caligiore 2022]. Economist Klaus Schwab, founder and president of the World Economic Forum,[3] argues that *combining biological, physical, and digital technologies* will lead to significant changes across all disciplines and sectors, challenging even our understanding of what it means to be human. This represents the *fourth industrial revolution*, which is already transforming how we live, work, and communicate [Schwab 2016].

The medical field will be among the most impacted by the fourth industrial revolution and the influence of AI. Advanced AI technologies have been enhancing and supporting medicine for several years. For instance, robots assist doctors with precision microsurgery and enable remote health monitoring through telemedicine. AI algorithms could also enhance the quality of radiological images, such as those from MRI scans. While these early AI applications in healthcare represent modernization, they have not fundamentally transformed medical practice. However, in a few years, the landscape will be drastically different. The convergence of digital and information technologies, neurotechnologies, and AI will drive a true Copernican medical revolution. Terms like *healthcare 2.0* or *e-Health* often describe this profound shift [Eysenbach 2001]. The increasing digitization[4] of patient data from diverse sources – such as electronic medical records, public research databases, clinical trials, smartphones, apps, and wearable health devices – will be critical for unleashing the full potential of AI algorithms, which rely on digital data to learn and solve problems.

The AI revolution in medicine will hinge on an interdisciplinary approach and a holistic view of disease study. These elements will be pivotal in advancing precision and personalized medicine. Collaboration among experts from diverse fields brings together varied perspectives, knowledge, and expertise, fostering a deeper understanding of the biological processes behind diseases and encouraging innovative approaches in medical research. An interdisciplinary approach to disease prevention and treatment considers multiple factors, including social and economic conditions, lifestyle, environment, and individual genetic variability. The aim is to develop tailored treatments for each patient, enhancing treatment effectiveness and optimizing clinical outcomes [Collins et al. 2017]. AI will enable the processing and analysis of vast amounts of diverse patient data – typically examined separately by different disciplines, such as genetic, behavioral, physiological, demographic, and psychological data – to identify new indicators that assist physicians in making faster and more accurate diagnoses. AI can also predict the likelihood of developing certain diseases. This level of analysis is beyond the capability of traditional statistical methods. In recent years, scientific publications have significantly increased, proposing machine learning algorithms that can aid physicians in diagnosing various diseases by processing large, complex datasets [Myszczynska et al. 2020].

AI will act as a magnifying glass for medicine, enabling doctors to observe the human body with greater detail and precision. Like a microscope detecting damaged cells or a map highlighting brain activity during specific tasks, AI can uncover valuable insights that might go unnoticed, aiding in more accurate and personalized diagnoses and treatments. A key element of AI in medicine will be its ability to address illnesses by considering the unique characteristics of everyone, such as genetic factors and lifestyle, while simultaneously considering the complex system that makes up the person. This system includes the interconnected aspects of the individual brain, body, and environment. AI could gather information by integrating diverse data sources, such as genetic data, to identify potential predispositions to specific diseases or data from wearable devices and environmental sensors on how surroundings, lifestyle, diet, physical activity, and sleep quality affect the patient. By analyzing this wealth of information, AI could reveal correlations between genetic factors, lifestyle, and environment, offering a 'system view' of the patient's health. Moreover, AI can assist physicians in personalizing treatments with greater precision by considering the typical symptoms of a disease and the patient's unique risk factors. Additionally, it can recommend preventive, lifestyle-based interventions to improve overall health.

Doctors and researchers will use AI to create digital twins of patients, capable of simulating the functioning of a person's body and brain on a computer. These digital twins will be critical in advancing predictive and personalized medicine. By virtually simulating the effects of various therapies for specific patients, they can help doctors make more informed decisions. Digital twin simulations will also enable the investigation of the underlying causes of diseases. Thus, AI will profoundly transform how we receive medical care, reshape the doctor-patient relationship, and revolutionize scientific research by deepening our understanding of disease mechanisms, exploring links between different conditions, and discovering new treatments.

All this will bring enormous benefits, but it will also introduce new challenges that will be difficult to address as they involve unprecedented issues. For example, when a doctor relies on AI to assist in patient treatment decisions, who bears responsibility for those choices? Is it the doctor, the AI itself, the developers, or the manufacturers? How should these ethical and legal issues be managed? How can we prevent emerging technologies that combine AI and neurotechnologies from manipulating the body and mind? Will AI eventually replace doctors and healthcare professionals? Leveraging the most recent scientific research, this book explores the most innovative applications of AI in medicine, offering a clear and multidisciplinary perspective on how these advancements will forever change healthcare. It also provides an overview of the new challenges posed by AI rise in the medical field and proposes possible solutions for its ethical and sustainable development.

NOTES

1. Neurotechnology combines knowledge from neuroscience with fields like informatics, mathematics, psychology, artificial intelligence, and robotics to develop devices that interpret brain signals and use them to interact with the body or environment. These tools enable the study of the nervous system, improve the diagnosis and prevention of neurological disorders, and in some cases, enhance therapies to alleviate symptoms associated with such conditions.
2. Information technology (IT) refers to systems capable of storing, processing, and transmitting data through communication networks. These networks, which can be wired, optical (fiber-optic), or wireless, facilitate the exchange of information between users in different locations. The internet is a prime example of a communication network that has driven the advancement of IT. Widely utilized across all professional, economic, and social sectors, IT is often associated with digital technologies – electronic devices like computers, smartphones, and tablets that process data using a binary system of two digits: 0 and 1.
3. The World Economic Forum is an international organization that fosters collaboration between the public and private sectors to improve the world state. Each year, the Forum hosts a meeting in Switzerland, bringing together prominent figures from global politics, business, academia, and the media to discuss the world most pressing challenges. The Forum also regularly publishes a series of research reports.
4. Digitization consists of converting information from images, text, sound, video, or any analog signal that can take on a continuous range of values into a digital format. A binary code represents digital data through combinations of two digits: 0 and 1.

REFERENCES

Caligiore, D. [2022], *IA istruzioni per l'uso*, Bologna, Il Mulino.

Collins, D. C., Sundar, R., Lim, J. S. & e Yap, T. A. [2017], *Towards Precision Medicine in the Clinic: From Biomarker Discovery to Novel Therapeutics*, Trends in Pharmacological Sciences, 38, n. 1, pp. 25–40.

Eysenbach, G. [2001], *What Is E-Health? Journal of Medical Internet Research*, 3, n. 2, p. e833.

Myszczynska, M. A., Ojamies, P. N., Lacoste, A. M., Neil, D., Saffari, A., Mead, R. ... & e Ferraiuolo, L. [2020] *Applications of Machine Learning to Diagnosis and Treatment of Neurodegenerative Diseases*, Nature Reviews Neurology, 16, n. 8, pp. 440–456.

Schwab, K. [2016], *La Quarta Rivoluzione Industriale*, Milano, FrancoAngeli.

Acknowledgments

I express my gratitude to all those who, directly or indirectly, inspired many of the ideas discussed in this book. Thanks to Gianfrancesco Angelini, Samuele Carli, Fabio Massimo D'Amore, Lorenzo Morena, Marco Moscatelli, Simone Torsello, my colleagues at the Italian National Research Council – Institute of Cognitive Sciences and Technologies and AI2Life srl, and the lecturers and students at the Advanced School in Artificial Intelligence. I would also like to express my gratitude to Marta Mazzucchelli from 'Il Mulino' and Elliott Morsia from 'Routledge Taylor & Francis' for their valuable support throughout the editorial process. My heartfelt thanks go to Serena, Chiara, Carla, Claudio, Francesca, Maria Enza, Mauro, Miriam, Mirco, Nella, Nello, Paola, and Salvatore for their support and encouragement.

CHAPTER 1

What Artificial Intelligence Is and Its Historical Applications in Medicine

1.1 ARTIFICIAL INTELLIGENCE, MACHINE LEARNING, AND DEEP LEARNING

Artificial intelligence focuses on developing computer programs that can analyze situations and *learn* from experience to solve problems autonomously, without human intervention. Throughout this text, we will refer to artificial intelligence simply as AI. As a highly multidisciplinary field, AI vast impact and influence across numerous areas mean that not only engineers, mathematicians, and computer scientists are involved, but also psychologists, philosophers, lawyers, neuroscientists, and scholars from various disciplines who are interested in its ethical, theoretical, and practical implications. The purpose of AI can be either technological, aimed at solving practical problems, or scientific, used to explore questions related to human beings, other living organisms, or nature.

Not all AI systems have learning capabilities. Some are designed to perform specific, often repetitive tasks based on human instructions and do not require specialized training. AI development approaches fall into two

DOI: 10.1201/9781003606130-1

broad categories: The symbolic approach, which does not involve learning, and the subsymbolic approach, which does.

AI algorithms that use the symbolic approach are, for example, the expert systems. They do not learn how to solve problems; instead, they operate by incorporating a series of sequential instructions (algorithms[1]) into the AI program. These instructions contain knowledge from human experts in a specific field, allowing the system to address problems based on predefined logic. For instance, an AI system applied in medicine might follow steps like the ones below.

1. Collecting individual patient data: Gather the patient's medical information, including symptoms, lab results, and medical history.[2]

2. Data preprocessing: Clean and normalize the data (e.g., transform all numerical values to a scale of 0–1)[3] to remove errors and inconsistencies. Convert raw data into a format suitable for analysis, such as organizing it into a table (matrix).

3. Algorithm logic construction: Develop the software using predefined rules from medical experts for a specific pathology. These rules, including cause-and-effect relationships, diagnostic criteria, and treatment guidelines, form the instructions (logic) that the symbolic AI follows. Use logic programming languages or symbolic representations to encode this knowledge. For example: "IF the patient has a high fever and abdominal pain, THEN they might have a gastrointestinal infection".

4. Analyzing patient data with the algorithm: Apply the algorithm to the patient's data using the established medical rules. This process generates a list of possible diagnoses based on the rules from step 3. The algorithm could assign a probability score to each diagnosis, aiding the physician in making informed decisions or recommending further tests or treatments.

A symbolic AI system in medicine relies on expert-developed rules to analyze patient data, offering possible diagnoses or recommendations that assist doctors in decision-making. Expert systems operate in specific professional fields, detecting anomalies in data and identifying potential causes, which makes them particularly useful for diagnostics and monitoring critical conditions. They also help plan sequences of actions to achieve specific goals within set constraints. For example, a doctor treating a patient with diabetes can use an expert system to plan an optimal treatment regimen, considering various factors. The system analyzes the patient's data, such as blood glucose levels, age, weight, physical activity, and other pre-existing conditions. Building on this information suggests a personalized course of action, such as tailored dietary plans, insulin dosages, exercise routines, and regular blood sugar monitoring.

The symbolic approach is a *top-down* process. It begins with expert knowledge (*the knowledge level*), which is then converted into a system of symbols (*the symbol level*). These symbols are further transformed into binary code (*the numeric level*) that a computer can process. For example, if medical experts know that a person with a fever and a stuffy nose might have a cold, this constitutes the knowledge level. We then translate this into symbolic form using instructions like 'IF (fever and stuffy nose) THEN (cold)'. Here, 'fever' and 'stuffy nose' are symbols representing symptoms, while 'cold' is the result if both symptoms are present. Next, we convert these symbols into binary code for computer processing. For instance, we might represent the presence of 'fever' with 1, its absence with 0, and do the same for 'stuffy nose' and 'cold'. If both symptoms are present, the system will output 1 at the numeric level, indicating the presence of a cold.

AI systems using the symbolic approach offer several advantages. They are highly effective for making rapid and accurate decisions related to specific tasks, such as diagnosing a particular disease in medical applications. Compared to human domain experts, these systems can solve the same problem more quickly, with fewer errors, and without fatigue. Additionally, the logic behind the AI decision-making process is transparent and can be traced through the chain of 'IF' and 'THEN' statements programmed by human experts. However, these systems also have significant drawbacks. They cannot learn autonomously and are suitable for particular domains, which limits their flexibility for other tasks. They require manual updates from programmers and domain experts to address environmental variability. Furthermore, these systems could have high costs for development and maintenance.

The subsymbolic approach, also known as the 'mathematical-statistical' or 'numerical' approach, forms the basis of learning AI or 'machine learning' (ML) systems. Unlike symbolic systems, where human experts program the solution into AI, ML models learn to solve problems by analyzing examples from data. However, this does not diminish the importance of medical experts in ML algorithm development. Physicians play a crucial role in various stages, such as obtaining and processing data for training the ML model, interpreting the model results, and integrating it into the medical environment to ensure its effective and safe use. The difference is that, in the subsymbolic approach, the algorithm itself uncovers the logic required to solve the problem, rather than having physicians explicitly encode this logic as in symbolic AI systems.

The subsymbolic approach follows a *bottom-up* process, starting with raw data input at the numerical level, which the computer then processes. The ML model then autonomously learns from the data to extract knowledge. For example, to develop a subsymbolic AI system that assists physicians in diagnosing lung diseases by analyzing X-ray images, the process could be implemented through the following steps.

1. Collect data from many subjects: Gather a substantial dataset of chest X-rays from various individuals. Medical experts must carefully annotate each image, labeling it with the corresponding disease. Typically, the amount of data needed increases with the complexity of the problem the ML model aims to solve.

2. Data preprocessing: Standardize the size of X-ray images, reduce noise, and enhance image quality. Additional steps may include normalizing the data, addressing missing values, and removing outliers.[4]

3. Feature extraction: Apply techniques to convert X-ray images into tabular data, also known as features.[5] Features are the key characteristics that define and distinguish a subject from others. Just as each person has a unique set of features that sets them apart, so do objects like houses, flowers, and images. In a dataset of chest X-rays, relevant features might include the image height and width, which offer information about the size of the chest or the X-rayed region, pixel density for detecting areas of interest or abnormalities like lung opacities, and the edges and contours of chest structures, such as ribs.

4. Creating the training and test datasets: Split the dataset into two parts, one for training the model and the other for evaluating its performance. A common approach is to allocate 80% of the data for training and 20% for testing.

5. Selecting the ML algorithm: Choose a suitable ML algorithm for classifying X-ray images. Deep neural networks, particularly Convolutional Neural Networks (CNNs), are often preferred for computer vision tasks. CNNs automatically learn relevant features

directly from the images, reducing the need for manual feature extraction. However, unlike some ML algorithms that rely on manual feature extraction, CNNs may offer less transparency in interpreting radiographic images, leading to significant concerns about the explainability of their results (see Chapter 3, Section 3.5).

6. Training the model: Utilize the training dataset to teach the ML model. In this phase, the model learns to identify lung disease signs from radiographic images and tabular data.

7. Optimizing the model involves adjusting its parameters to enhance performance. This process can be time-consuming.

8. Model evaluation: Assess the model performance using the test dataset, which consists of data the algorithm did not encounter during training. Measure metrics such as accuracy, sensitivity, specificity, and precision[6] to evaluate how effectively the model recognizes lung disease.

Once an ML algorithm completes training, it can provide solutions through the 'inference' process. During inference, the model applies what it has learned to new, unseen data to generate predictions or answers related to the problem the model has to solve. The goal is to extend the model predictive capabilities beyond the data used during training. A well-designed model generalizes effectively, delivering accurate results on fresh data (this is known as the *generalization* property). For example, imagine a doctor analyzing chest X-rays using a subsymbolic AI system built following the previously outlined steps. After training on thousands of X-ray images, the AI has learned to recognize signs of lung disease. The doctor loads the patient's X-ray into the system. The AI identifies suspicious or diseased areas and suggests follow-up actions, like conducting further tests or biopsies in specific parts of the lungs. These recommendations assist the doctor in making informed decisions, enabling earlier diagnoses and more effective treatments. In this scenario, the subsymbolic AI acts as a virtual assistant, aiding the physician in interpreting medical images. This support can lead to earlier diagnoses and more effective treatments for patients.

Like the symbolic approach, ML models facilitate quick and accurate decision-making, often reducing error rates compared to human operators. The subsymbolic approach, however, offers distinct advantages. For example, the same model might address different problems by changing the training data. Additionally, these models autonomously learn the problem-solving logic from the data, which allows them to adapt to changing conditions without human expert intervention, ultimately lowering development and update costs. Despite these benefits, the subsymbolic approach has notable drawbacks. It is often challenging to trace the logic behind the algorithm decisions (the explainability problem, see Chapter 3, Section 3.5), as the numerical representations used in the model processing layers may not align with human logic. Furthermore, solving complex problems might require large datasets, significant computational resources, and prolonged training times.

Unlike symbolic systems, where the algorithm structure – defined by the human expert (e.g., the IF/THEN rules) – determines its problem-solving specificity, subsymbolic systems rely on the training dataset to establish the algorithm specialization. In subsymbolic systems, the ML model structure remains flexible and adaptable to different problems based on the data used for training (this is known as the *flexibility* property). For example, CNN initially trained to detect lung diseases in X-ray images can be adapted to analyze skin images for dermatological purposes, provided it undergoes retraining with the appropriate data. Today, rather than competing, symbolic and subsymbolic AI techniques are increasingly collaborating to leverage the strengths of both approaches.

ML algorithms fall into two broad categories based on their problem-solving approaches: *Algebraic-statistical models* and artificial neural networks. Algebraic-statistical models leverage mathematical and statistical principles to analyze and interpret data, employing equations, formulas, and algorithms to uncover relationships within the training data. These models often perform well, even with limited data. In contrast, *artificial neural networks* take inspiration from biological neural networks. They feature interconnected 'artificial neurons' that mimic the connections between nerve cells in biological systems. These neurons use mathematical equations to exchange information and learn to solve tasks based on the data they receive. After training, neural networks can predict outcomes or classify data. They excel in processing structured data, such as numerical tables, and unstructured data, like text, images, audio, and video. Deep neural networks can analyze and generate text, identify objects in images, and understand spoken language. On the other hand, algebraic-statistical algorithms, such as decision trees and linear regression,[7] generally work better with structured data and face challenges when handling unstructured data efficiently and accurately [Boden 2019; Caligiore 2022].

ML algorithms perform three main functions: Classification, prediction, and clustering. In *classification*, the algorithm sorts objects into categories based on specific characteristics. For example, it can classify patients as 'sick' or 'healthy' by analyzing their symptoms and clinical test results, helping to distinguish between different health conditions. *Prediction* involves using regression to estimate numerical values based on historical data. For instance, an algorithm might predict how long it will take a patient to recover from surgery by examining recovery data from other patients, providing estimates for future outcomes. *Clustering* groups objects based on similarities without predefined labels. When working with a dataset that lacks explicit solutions, clustering identifies patterns by grouping patients with similar test results without assigning a specific diagnosis. This approach reveals common traits and patterns among different patient groups. In medicine, classification, prediction, and clustering support various tasks, including diagnosing diseases, forecasting treatment outcomes, and discovering patterns in patient data for research. Box 1.1 outlines the learning techniques to train ML models for classification, prediction, and clustering.

BOX 1.1 LEARNING PROCESSES FOR TRAINING MACHINE LEARNING SYSTEMS

Supervised learning. It is a widely used technique in the medical field when working with datasets that contain the solutions to specific problems, known as labeled data. This type of learning is crucial for training algorithms to perform tasks such as diagnosing diseases. For instance, the ML-based diagnostic system discussed earlier, which employs CNN to identify the presence or absence of a disease based on medical images, utilizes supervised learning. In this scenario, the training dataset contains X-rays labeled "positive" for disease presence and "negative" for absence. The supervised learning algorithm begins training by analyzing the X-rays and comparing its predictions to the provided labels. During the initial stages, the algorithm will inevitably make errors – such as misclassifying a negative X-ray as positive or vice versa. These errors help the algorithm optimize its parameters. In particular, the algorithm aims to gradually reduce these errors by adjusting its responses to align more accurately with the training labels. This iterative process continues until the algorithm achieves a satisfactory diagnostic accuracy level.

Unsupervised learning. It applies when the training dataset lacks explicit information about the answers the AI should provide. In this context, unsupervised learning algorithms are crucial for detecting implicit correlations in the data without the need for manually labeled outputs. To illustrate its usefulness in medicine, consider a hospital that collects patient data, such as diagnostic images, laboratory test results, and medical histories, without explicitly indicating whether patients have specific pathologies. In this scenario, unsupervised learning can organize patients into clusters based on shared characteristics, revealing hidden correlations among various factors. For instance, unsupervised learning might identify and put in the same cluster patients with common traits – such as age, family history, and clinical test results – who exhibit similar trends in symptoms or diagnoses. This insight could improve the understanding of how risk factors relate to diseases, helping physicians identify vulnerable patients more effectively.

Reinforcement learning. Unlike supervised learning, reinforcement learning does not provide a direct problem solution. Instead, the system receives positive or negative feedback (reward) when it achieves specific goals defined by a human operator. The machine learning algorithm continuously adapts and optimizes its parameters to maximize these rewards during interactions with the environment. An example of this technique in the medical field is managing drug administration for patients in intensive care. Maintaining drug levels within therapeutic ranges is crucial for effective and safe treatment, yet these levels can fluctuate based on a patient's condition, necessitating constant monitoring and dosage adjustments. A reinforcement learning system could help automate and optimize this process. In this scenario, the learning environment encompasses the intensive care unit, the patient, monitors for vital signs such as blood pressure and heart rate, and the drug infusion system. Initially, the system may lack knowledge about the optimal dosages for a specific patient and might propose actions, such as adjusting the infusion rate of a particular drug. Based on the patient's response, the system receives feedback: A positive reward if the patient's vital signs improve and drug levels remain within the therapeutic range, or a negative reward if drug levels are inappropriate and the patient's condition worsens. During the early stages, the physician typically evaluates the AI-suggested actions without implementing them directly, thus preventing potential harm in cases of negative reinforcement. The doctor uses their experience to determine whether the feedback is positive or negative. Over time, the algorithm should enhance its ability to recommend optimal drug adjustments, helping to keep the patient in a desired therapeutic state. This reinforcement learning approach allows for continuous monitoring and adjustment of drug administration tailored to the patient's needs, reducing the workload for medical staff and improving treatment accuracy.

This book primarily focuses on the subsymbolic approach, which is revolutionizing our daily lives. Understanding the distinction between symbolic and subsymbolic AI is essential for grasping the differences between AI, ML, and deep learning. While these terms are often used interchangeably, this is not entirely accurate. The main objective of AI is to create machines – whether physical devices or software simulations – that exhibit human-like intelligence. Therefore, AI aims to replicate various characteristics of human intellect, such as planning, language comprehension, object recognition and manipulation, and problem-solving. The term "AI" encompasses all forms of artificial intelligence, including those that learn (subsymbolic AI) and those that do not (symbolic AI). The term "ML" emerged later, signifying a machine ability to learn without explicit programming. ML represents a subset of AI, focusing on how machines can autonomously learn tasks from a given dataset. Deep learning, on the other hand, is an ML technique that utilizes complex artificial neural networks with multiple processing layers, known as deep neural networks. The term "deep" refers to the multiple layers of data processing involved; much like the human brain, deep learning models consist of several layers of neurons between the input and output layers. A key distinction between ML and deep learning lies in how these systems perform relative to the data used for training. Deep learning models tend to improve as the quantity and quality of training data increase, which is a fundamental property driving their growing success. We are currently in an era marked by an extraordinary surge in digital data production, sourced from mobile devices and the online environment, primarily managed by tech giants like Google, Amazon, and Facebook. Additionally, deep learning models excel at analyzing unstructured data such as video, audio, and text. Significant advancements in processing speed and computing power have made training deep neural networks with multiple layers faster than ever. All these factors contribute to the success of deep learning. In summary, AI encompasses ML and various other techniques, including symbolic systems. Within this framework, ML includes deep learning along with other approaches often grounded in statistics and probability theory.

1.2 THE IMPORTANCE OF DIGITISING DATA

The digitization of patient data is essential for advancing the future of medicine for several reasons. First, it allows patient information to be stored electronically, enabling the creation of electronic health records (EHRs). An EHR is a digital version of the traditional paper-based medical document used to monitor and manage an individual's health information. It can contain a patient's complete medical history, e.g., treatments, allergies, and lab results. With EHRs, the need to manually sift through paper records is eliminated, significantly speeding up access to critical patient information. Additionally, EHRs make patient data readily available to healthcare professionals across different locations and in real-time, improving information sharing between hospitals, clinics, general practitioners, and specialists. This seamless access enhances the quality of care and promotes a more effective, coordinated healthcare system.

The contamination between AI and digital and information technologies could be critical for extracting valuable insights from EHRs and optimizing administrative processes. Often, doctors spend more time filling out EHRs than interacting with patients. However, AI algorithms could automate the compilation of EHRs [Chen et al. 2018], reducing the administrative burden on physicians. This approach would allow them to focus on utilizing the information within the records and fully benefit from the advantages of digital tools, such as seamless information sharing, ultimately giving them more time to dedicate to patient care [Reitz et al. 2012].

Digitizing patient data is crucial for harnessing the full potential of AI algorithms, which learn from examples provided in a digital format. This process is critical for advancing *personalized medicine*, an innovative approach to disease prevention and treatment that tailors therapies to each patient by considering, for example, factors such as environment and genetic variability [Collins et al. 2017]. With this capability, AI algorithms can analyze and integrate large, diverse datasets, including genomic, clinical, and lifestyle information, to predict the likelihood of developing specific diseases. Additionally, AI can simulate the effects of different therapies virtually, allowing for predictions about the success rates of various treatments (see Chapters 2 and 4).

The digitization of data and the application of AI can greatly enhance the healthcare system efficiency. These advancements can lead to reduced administrative costs, optimized resource management, and more effective planning of healthcare services. For instance, software can automate the booking process for healthcare services and assist patients through virtual assistants and chatbots. Moreover, AI techniques can extract valuable information from combined data, playing a crucial role in the command center operation. Such software collects and visualizes real-time data from all hospital operating systems on a monitor. Within this framework, AI algorithms can optimize hospital efficiency by providing recommendations for staff and detecting hospital-acquired infections. The examples presented in this chapter and Chapter 2 illustrate how digitizing patient data lays the groundwork for more efficient use of hospital information. This foundation is essential to fully harness AI potential in the medical field, improving diagnosis, treatment, research, and disease prevention while increasing overall healthcare efficiency. However, it is vital to ensure the security and privacy of digitized data throughout this process (see Chapter 3, Section 3.4).

1.3 MEDICAL IMAGING, ROBOTIC SURGERY, AND TELEMEDICINE

Several centers worldwide are already implementing cutting-edge technologies integrating AI, digital tools, and neurotechnology to enhance and support physicians' work [Reddy et al. 2020]. AI algorithms analyze patient data to generate valuable insights that assist in therapy planning and enable faster and more accurate diagnoses. AI has been applied in medical imaging for years, offering precise and rapid analysis of X-rays, computed tomography (CT) scans, and magnetic resonance imaging (MRIs) [Nichols, Herbert Chan and Baker 2019]. These systems can detect subtle abnormalities and early disease signs that may be missed by the human eye, facilitating earlier diagnoses and timely treatments, which improve patient outcomes.

Beyond medical imaging, other established AI applications in healthcare include robotic surgery and telemedicine. The AI role in surgery began about 25 years ago with the introduction of automated endoscopic system for optimal positioning (AESOP), a robotic system for laparoscopic,[8] minimally invasive procedures [Unger, Unger and Bass 1994]. Since then, several robotic surgical systems have been developed [Jakopec et al. 2001; Kwoh et al. 1988; Stefano 2017]. A key driver in the evolution of robotic surgery was telepresence, a system developed through collaboration between NASA Ames Research Center and Stanford researchers. Initially designed for military use, it allowed surgeons to operate remotely on wounded soldiers. In 2001, the ZEUS robotic system, an AESOP evolution, enabled the first remote minimally invasive surgery (telesurgery) [Hockstein et al. 2007; Lane 2018; Marescaux et al. 2001]. Known as the 'Lindbergh operation', French surgeons in New York successfully performed a cholecystectomy on a patient in Strasbourg. Telesurgery, with its capacity for remote operations regardless of location, can enhance access to care in underserved communities and specialized environments such as military settings. However, this technology has not been fully implemented because robotic systems rely on wireless networks, which face challenges in connection speed and stability. Delays in transmitting commands, video, and audio could compromise patient safety. Moreover, addressing legal and economic issues is crucial, especially when coordinating procedures across different centers [Legeza et al. 2021].

The Da Vinci robotic surgery system is one of the most advanced and widely used today [Stefanelli et al. 2020]. In general, surgical robots enable doctors to perform highly precise and stable movements by minimizing the effects of human tremors, leading to greater accuracy during procedures. This precision reduces blood loss during surgery, enhancing safety. Additionally, these robots provide a high-definition, three-dimensional view inside the patient's body, offering surgeons detailed visibility of the area they are working on. Robotic laparoscopy typically requires smaller incisions compared to traditional surgery, which lowers postoperative pain, shortens recovery time, and decreases the risk of infection. Moreover, the smaller, more precise incisions result in less noticeable scars, improving aesthetic outcomes. These robotic systems offer numerous benefits for both surgeons and patients by increasing the precision, safety, and comfort of surgical procedures. However, the success of robotic laparoscopy still relies heavily on the surgeon's experience and skill.

Robotic surgery today is primarily associated with minimally invasive procedures but still relies heavily on human input and control. Some robotic systems have a degree of autonomy; for instance, certain gastrointestinal suturing robots can adjust the suturing process based on data from sensors attached to the patient's body. This is known as "weak automation", where technology can perform specific tasks automatically but cannot interpret problems or provide solutions at the level of human expertise. On the other hand, "strong automation" refers to systems that can respond to their environment, interpret data, learn, and perform tasks autonomously. In the future, surgical robots may become more autonomous with the integration of AI capable of learning. Such robots could be programmed to carry out standardized procedures with consistent precision, reducing the risk of human error. Despite this potential for increased autonomy, human oversight and control will remain essential. Surgeons must supervise the entire process, make key decisions, and intervene in emergencies.

Telesurgery is one aspect of telemedicine. This approach combines AI and information technology to enable remote patient monitoring [Cortellessa et al. 2018], improving the management of chronic diseases, which make up a large share of healthcare costs. For instance, telemedicine could help to monitor patients with conditions requiring continuous monitoring, such as diabetes or cardiovascular diseases. These patients must regularly track critical parameters, like blood glucose levels for diabetics, to minimize the risk of complications. Telemedicine enables the exchange of these data between the patient (at home, in pharmacies, or care facilities) and a monitoring station, where a healthcare provider interprets the data. Telemedicine also plays a crucial role in diagnosis by facilitating the sharing of diagnostic information, though it does not entirely replace physical examinations. While performing a complete remote diagnosis can be challenging, telemedicine can offer valuable support by providing insights into the diagnostic process. In November 2022, Amazon introduced Amazon Clinic, a telemedicine system that allows patients to connect virtually with healthcare professionals for assistance with common medical concerns, such as allergies or migraines. After completing a brief medical history questionnaire, patients can consult with a doctor via video or text. The system aims to provide 24-hour coverage, ensuring timely assistance for patients.

1.4 EXOSKELETONS AND CYBORGS

In recent years, the strong synergy between AI, mechatronic systems,[9] and virtual and augmented reality devices has led to the development of innovative technologies in biomedical robotics for physical and mental rehabilitation, often called *phygital rehabilitation*. Advanced AI systems, combined with interactive environments and state-of-the-art technologies like *wearable robotics*, actively engage individuals in activities designed to enhance their functional abilities. An example of such wearable robotic devices is assistive exoskeletons – mechatronic orthoses that guide a person's limb movement and facilitate posture changes (see Figure 1.1). These versatile systems serve various purposes; in hazardous work environments, exoskeletons primarily aim to reduce worker fatigue and enhance safety. In the medical field, they assist individuals in daily activities and help rehabilitate the motor skills of patients who have experienced varying degrees of physical or neurological impairments. Exoskeletons can support mobility by aiding patients in walking, climbing stairs, and lifting heavy objects. They also enable individuals with muscular or skeletal

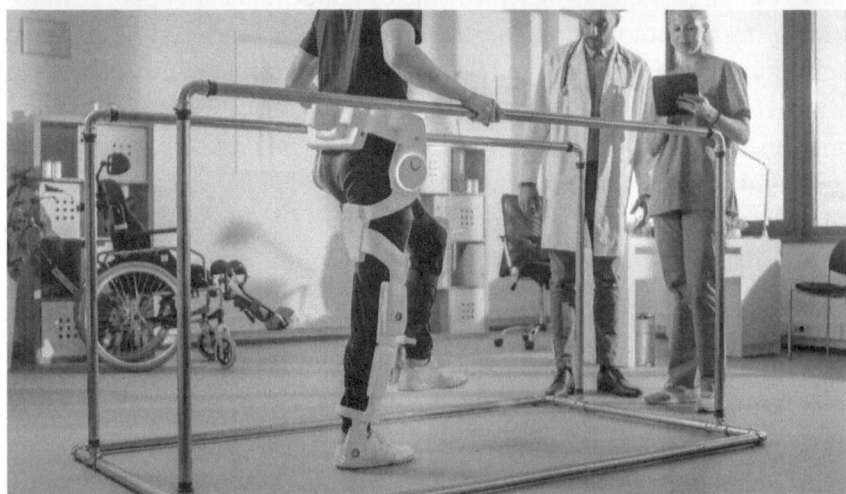

FIGURE 1.1   Possible applications of exoskeleton.

Source: (a) High-tech Futuristic Warehouse Worker Wearing Advanced Stock Photo 1845773707 | Shutterstock; (b) Modern Hospital Physical Therapy Patient Injury Stock Photo 2029416683 | Shutterstock.

disorders to perform rehabilitation exercises and physical therapy, helping them regain muscle control and strength. AI could allow exoskeletons to interact with the rehabilitation patient and the therapist. It could provide support during exercises and monitor biomechanical parameters related to the patient's progress. The collection and real-time processing of data using AI enable the creation of customized therapy protocols. These can make the often lengthy and challenging recovery process more efficient.

There are lighter, less invasive devices called *exosuits*. Unlike traditional exoskeletons, exosuits focus less on providing mechanical force and more on enhancing movement efficiency. Often worn like regular clothing, such as a tracksuit, they help reduce muscle fatigue and improve endurance during prolonged activities like walking or lifting heavy loads. Instead of relying on bulky mechanical components, exosuits typically use *smart materials* that can change shape or stiffness in response to stimuli like heat or electricity. For example, a smart material might function as a soft actuator[10] that contracts or expands when a specific electrical voltage is applied. AI algorithms that learn to recognize and classify a patient's movements could be critical in making exosuits more adaptable and integrated with the individual (personalized medicine). These wearable devices may not be suitable for all patients. For instance, those with severe skin lesions may face increased risks. In addition, exoskeletons can be costly and require specialized training for correct use, such as teaching physiotherapists to adapt them to specific patient needs.

Exoskeletons are wearable external supports, but more invasive methods involve implanting electronic components inside the body to restore functionality lost due to illness or injury. The idea of augmenting human abilities with artificial elements is not new. Since the 1960s, the term "cyborg" (a combination of "cybernetics"[11] and "organism") has described entities that exist at the intersection of human and machine, where individuals integrate artificial parts such as mechanical prostheses or electronic components into their bodies. British engineer Kevin Warwick is one of the first cyborgs in history. In the 1990s, he implanted an electronic chip to open doors and control lights remotely. Later, he added a second chip that connected him to a computer. Today, the distinction between humans and cyborgs is becoming increasingly blurred, especially with advancements in prosthetic technologies and artificial organs. For example, someone with a pacemaker might already be considered a cyborg. Recently, a young Italian man who lost the use of his legs in a car accident was able to walk again thanks to intelligent spinal cord stimulation technology developed by the University Hospital of Lausanne. This system creates a digital bridge between the brain and spinal cord, bypassing the damaged area caused by the injury. Similarly, research in Switzerland, led by teams at EPFL (headed by neuroscientist Grégoire Courtine) and the University of Lausanne (headed by neurosurgeon Jocelyne Bloch), has demonstrated that electrode-based stimulation of the lower spinal cord can help individuals with partial paralysis regain their ability to walk.

NOTES

1. An algorithm is a sequence of well-defined instructions designed to perform a specific task or solve a particular problem. For instance, the algorithm for "preparing a tomato and cucumber salad" can be outlined in the following steps: (1) Gather the necessary ingredients: Tomatoes, cucumbers, olive oil, and salt; (2) wash the tomatoes and cucumbers thoroughly under running water; (3) slice the tomatoes into thin pieces and cut the cucumbers into rounds; (4) in a large bowl, combine the sliced tomatoes and cucumbers; (5) drizzle olive oil over the mixture and add a pinch of salt; (6) lightly mix everything so that the dressing gets evenly spread.

2. The patient's medical history includes details about past illnesses, allergies, surgeries, medications, and current symptoms. These data are crucial for accurate diagnosis and effective treatment planning. Ensuring the quality of medical history is essential for providing optimal patient care.

3. Normalization eliminates scale differences between diverse variables, facilitating better comparison and analysis. This process may involve transforming the data to achieve a mean of 0 and a standard deviation of 1 or scaling values to a specific range, typically between 0 and 1. By normalizing the data, analysis becomes more straightforward, particularly when working with heterogeneous datasets that exhibit a wide range of variation.

4. An outlier is a value that significantly deviates from most data points in a dataset. In other words, it is an unusual observation that diverges from the norm. Outliers can impact statistical analyses, so it is crucial to consider them carefully.

5. Features in a dataset are the characteristics or attributes that describe individual observations or samples. Essentially, they represent the specific information measured or recorded for each element in the dataset. For example, in a dataset of patients, features might include age, blood pressure, cholesterol levels, and more.

6. These values are crucial for evaluating ML algorithms. *Accuracy* measures the algorithm ability to make correct predictions; a higher accuracy indicates better performance. *Sensitivity* indicates the algorithm ability to detect true positives among all actual positives. True positives occur when the algorithm correctly identifies a positive result when it truly is positive, meaning the test accurately confirms the presence of the assessed condition. A high sensitivity value signifies that few true positives are missed. *Specificity*, on the other hand, assesses how well the algorithm identifies true negatives among all actual negatives. True negatives reflect situations where the algorithm correctly confirms the absence of the condition. A high specificity value indicates that few negative cases are misclassified. *Precision*

measures the reliability of the positive predictions made by the algorithm. A high precision value indicates that the positive predictions are mostly correct.

7. A *decision tree* is a model that makes decisions by asking a series of binary (yes/no) questions. It starts with a principal question and branches out based on the answers, forming a tree-like structure. Training a decision tree involves automatically learning to ask these questions optimally to make accurate decisions, enabling the model to classify inputs effectively. In contrast, *linear regression* is a model that identifies the best linear input-output variables relationship. It analyzes the data to find the line that best fits the points in a graph representing the input and output variables. Linear regression aims to understand how one variable linearly influences another.

8. In laparoscopic surgery, doctors operate within the human body using specialized instruments and a small camera. They create tiny incisions to insert these tools, allowing them to visualize and perform procedures without fully opening the affected area.

9. Mechatronic devices integrate the features of mechanical and electronic systems, with the term "mechatronics" stemming from the combination of "mechanics" and "electronics".

10. An actuator is a device that transforms a control signal into a physical movement or action.

11. Cybernetics is the science that explores control and communication systems in both humans and machines.

REFERENCES

Boden, M. [2019], *L'Intelligenza Artificiale*, Bologna, Il Mulino.

Caligiore, D. [2022], *IA istruzioni per l'uso*, Bologna, Il Mulino.

Chen, X., Liu, Z., Wei, L., Yan, J., Hao, T. & Ding, R. [2018], *A Comparative Quantitative Study of Utilizing Artificial Intelligence on Electronic Health Records in the USA and China During 2008–2017*, BMC Medical Informatics and Decision Making, 18, n. 5, pp. 55–69.

Collins, D. C., Sundar, R., Lim, J. S. & Yap, T. A. [2017], *Towards Precision Medicine in the Clinic: From Biomarker Discovery to Novel Therapeutics*, Trends in Pharmacological Sciences, 38, n. 1, pp. 25–40.

Cortellessa, G., Fracasso, F., Sorrentino, A., Orlandini, A., Bernardi, G., Coraci, L., … & Cesta, A. [2018], *ROBIN, a Telepresence Robot to Support Older Users Monitoring and Social Inclusion: Development and Evaluation*, Telemedicine and e-Health, 24, n. 2, pp. 145–154.

Hockstein, N. G., Gourin, C. G., Faust, R. A. & Terris, D. J. [2007], *A History of Robots: From Science Fiction to Surgical Robotics*, Journal of Robotic Surgery, 1, n. 2, pp. 113–118.

Jakopec, M., Harris, S. J., Rodriguez y Baena, F., Gomes, P., Cobb, J. & Davies, B. L. [2001], *The First Clinical Application of a "Hands-On" Robotic Knee Surgery System*, Computer Aided Surgery, 6, n. 6, pp. 329–339.

Kwoh, Y. S., Hou, J., Jonckheere, E. A. & Hayati, S. [1988], *A Robot with Improved Absolute Positioning Accuracy for CT Guided Stereotactic Brain Surgery*, IEEE Transactions on Biomedical Engineering, 35, n. 2, pp. 153–160.

Lane, T. [2018], *A Short History of Robotic Surgery*, The Annals of the Royal College of Surgeons of England, 100, n. 6_sup, pp. 5–7.

Legeza, P., Sconzert, K., Sungur, J. M., Loh, T. M., Britz, G. & Lumsden, A. [2021], *Preclinical Study Testing Feasibility and Technical Requirements for Successful Telerobotic Long Distance Peripheral Vascular Intervention*, The International Journal of Medical Robotics and Computer Assisted Surgery, 17, n. 3, p. e2249.

Marescaux, J., Leroy, J., Gagner, M., Rubino, F., Mutter, D., Vix, M., … & Smith, M. K. [2001], *Transatlantic Robot-Assisted Telesurgery*, Nature, 413, n. 6854, pp. 379–380.

Nichols, J. A., Herbert Chan, H. W. & Baker, M. A. [2019], *Machine Learning: Applications of Artificial Intelligence to Imaging and Diagnosis*, Biophysical Reviews, 11, pp. 111–118.

Reddy, S., Allan, S., Coghlan, S. & Cooper, P. [2020], *A Governance Model for the Application of AI in Health Care*, Journal of the American Medical Informatics Association, 27, n. 3, pp. 491–497.

Reitz, R., Common, K., Fifield, P. & Stiasny, E. [2012], *Collaboration in the Presence of an Electronic Health Record*, Families, Systems, & Health, 30, n. 1, pp. 72–80.

Stefanelli, L. V., Mandelaris, G. A., Franchina, A., Di Nardo, D., Galli, M., Pagliarulo, M., … & Gambarini, G. [2020], *Accuracy Evaluation of 14 Maxillary Full Arch Implant Treatments Performed with Da Vinci Bridge: A Case Series*, Materials, 13, n. 12, p. 2806.

Stefano, G. B. [2017], *Robotic Surgery: Fast Forward to Telemedicine*, Medical Science Monitor, 23, p. 1856.

Unger, S. W., Unger, H. M. & Bass, R. T. [1994], *AESOP Robotic Arm*, Surgical Endoscopy, 8, p. 1131.

CHAPTER 2

How AI Will Revolutionize Disease Diagnosis, Treatment, and Prevention

2.1 DATA-DRIVEN AI AND THEORY-DRIVEN AI

There are two main approaches to using artificial intelligence (AI): The *data-driven* and the *theory-driven* approaches. The data-driven technique relies on data to identify correlations and patterns. Machine learning, a critical part of this approach, helps develop predictive, classification, or clustering models (see Chapter 1, Section 1.1) for diagnosing and treating diseases. For example, researchers can use machine learning to identify common risk factors for developing specific diseases. With the growing digitization of data and the powerful computing and storage capabilities of modern computers, it has become possible to efficiently collect and analyze vast amounts of data on human behavior and brain activity. AI algorithms, including deep neural networks and machine learning models, allow data integration from multiple sources, such as behavioral, physiological, and brain imaging studies. This data can come from experiments that examine various body and brain functions, such as how the heart responds to stress or how the brain handles cognitive tasks like memory and decision-making. The data-driven method aims to analyze data using AI to uncover

DOI: 10.1201/9781003606130-2

correlations, predict new behavioral patterns, and identify important factors in disease development. Recently, an international research group, including several Italian labs, applied a data-driven approach to developing a machine learning algorithm capable of predicting, within a three-year window, the progression to Alzheimer's disease in individuals with mild cognitive impairment (MCI) [Grassi et al. 2019]. The researchers trained the AI model using a wide range of easily accessible data, including demographic details and results from standard neuropsychological tests. This approach is generally valuable for diagnosing disorders when prior knowledge of the disease is limited or uncertain.

Machine learning is becoming increasingly important in medicine, allowing researchers to create mathematical models and software that help explore disease mechanisms, diagnose and treat illnesses, and even prevent certain conditions. Economically, the market for AI in healthcare is experiencing rapid growth, with forecasts predicting an increase in value from around USD 11 billion in 2021 to nearly USD 188 billion by 2030, reflecting an annual growth rate of 37%. This growth is driven by the numerous benefits of integrating AI into healthcare.[1] Additionally, a significant indicator of the impact of machine learning in medicine is the exponential rise in scientific research utilizing ML in the medical field in recent years (see Figure 2.1).

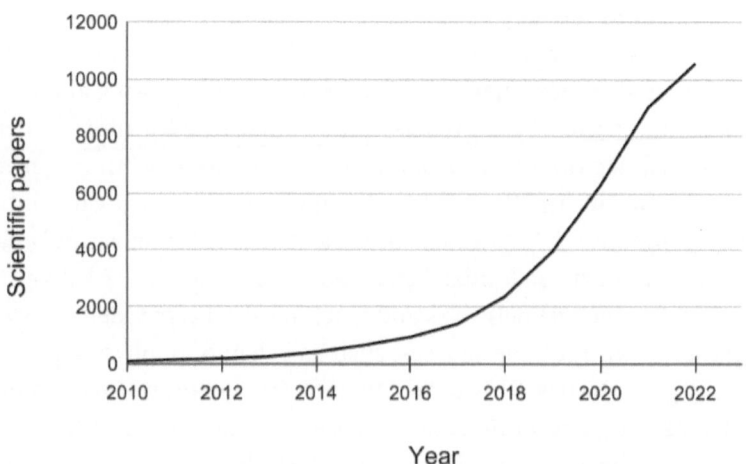

FIGURE 2.1 Growth in scientific publications on the use of machine learning (data-driven approach) in medicine in recent years. Data retrieved from PubMed[2] using "machine learning medicine" as the search keywords.

The theory-driven approach relies on existing knowledge of medical phenomena, utilizing established models and theories to develop AI systems that simulate bodily functions. This method enables AI to confirm or refute pre-existing medical hypotheses, enhancing researchers understanding of diseases and their underlying mechanisms. For instance, a researcher might employ AI to investigate whether inflammation is a causal factor in the progression of a specific disease. In this approach, empirical data form the basis for developing a theory about the physiological processes that underlie a phenomenon. To assess the validity of this theory, researchers develop a computational model (digital twin) that incorporates the initial hypothesis. For example, in a brain model, Area A connects to Area B through excitatory links and to Area C through inhibitory ones. This digital twin is then used to replicate experimental data collected from real subjects, helping to determine if the physiological processes it simulates can explain observed behaviors in humans. The theory-driven model can generate new data by predicting results that can be tested in future experiments. It also provides insight into the physiological mechanisms behind diseases. This understanding is valuable for designing early diagnostic methods or testing the effects of potential new therapies through computer simulations, using the theory-driven model as a digital twin of the patient.

It is worth noting that these two approaches can complement each other. For example, anatomical and functional assumptions (from the theory-driven approach) can help filter the vast amounts of data used to train ML algorithms (data-driven approach), thus aiding in the selection of significant data [Huys, Maia and Frank 2016]. Conversely, researchers can apply the data-driven method to analyze outputs from the theory-driven model to identify key parameters that characterize its functioning (*feature importance*) [Caligiore et al. 2017a].

Developing a theory-driven AI model to study, for example, the brain, requires *interdisciplinary expertise* in neuroscience, biology, computer science, engineering, mathematics, and psychology. Over time, those dedicated to building such models gain cross-disciplinary skills, enabling them to carefully select the most relevant computational elements for the mathematical equations underlying the model. Despite acquiring these skills, collaboration with disease-specific experts remains crucial for refining the model to address the specific problem and accurately interpreting the results. For example, when developing an AI model to study Alzheimer's disease, it is critical to involve neurobiologists and neuroscientists with specialized knowledge of the condition. The model serves as a bridge, fostering interdisciplinary collaboration and integrating different perspectives and approaches.

The theory-driven AI model could function as a *system-level model*. Developing a theory-driven model to study a neurodegenerative disease enables the simulation of brain regions typically linked to the condition while also considering their interactions with other relevant brain areas and connections to the rest of the body. The interdisciplinary and multi-methodological discussions that arise from creating a theory-driven AI system-level model are particularly compelling due to their potential to generate innovative solutions to complex problems. Researchers often struggle to achieve such solutions when focusing solely on a single area instead of considering the entire system. For example, experts studying Parkinson's disease typically concentrate on specific brain regions linked to the disorder, such as the basal ganglia – a subcortical area that regulates motor functions – and the dopaminergic system, which includes the substantia nigra, responsible for dopamine release. Dopamine is a neuromodulator[3] that influences various physiological and behavioral functions, including mood, motivation, movement, cognitive abilities, and sleep. The degeneration of neurons in the substantia nigra results in dopamine deficiency, a hallmark of Parkinson's disease. Consequently, these brain regions have received significant attention in understanding the mechanisms of the disease and in developing diagnostic and therapeutic interventions. However, it is crucial to note that the basal ganglia and the dopaminergic system do not operate in isolation. Recent research emphasizes the importance of examining significant interactions between these regions and other brain areas, such as the cerebellum and certain regions of the cortex. Moreover, other neuromodulators play a role in the early stages of Parkinson's disease development, even before obvious symptoms emerge [Caligiore, Giocondo and Silvetti 2022]. One of these neuromodulators is serotonin; its impairment seems to begin long before the onset of typical Parkinson's motor symptoms, such as tremors [Caligiore et al. 2021; Wilson et al. 2019]. Utilizing a theory-driven AI system-level model to examine interactions between the basal ganglia and other brain regions, as well as to investigate the role of neuromodulators beyond dopamine, could prove crucial in revolutionizing research on the causes and development of this disease. This approach could lead to new diagnostic methodologies and innovative treatments that extend beyond targeting just the basal ganglia and the dopaminergic system [Caligiore et al. 2017b].

Overall, these multidisciplinary and multimodal approaches could significantly impact the evolution of understanding of Parkinson's disease and, consequently, contribute to the discovery of more effective therapeutic solutions [Caligiore et al. 2016; Helmich 2018].

Traditional research often relies on single-disciplinary and specialized approaches that focus on specific areas of the body to study diseases, making it challenging to adopt a systemic perspective. Theory-driven system-level models enable researchers to analyze the brain and body by simulating their interactions on a computer [Parisi 2001]. This approach integrates data from diverse sources – such as clinical, behavioral, and brain imaging data – into a unified model, allowing simulations of new therapies on an artificial system. This method reduces reliance on living subjects for experimentation, providing economic and ethical advantages (see Chapter 4, Section 4.3).

Data-driven and theory-driven approaches extend beyond the medical field and apply to various phenomena. For example, data-driven AI could analyze traffic patterns, forecast market trends, and detect anomalies in computer systems. Similarly, theory-driven AI could aid in developing models for urban planning, analyzing climate change, and optimizing production processes. The following paragraphs provide several specific examples of data-driven and theory-driven models that could be developed to diagnose and treat various diseases.

2.2 DIGITAL TWINS, METAVERSE, AND ORGANOIDS

Digital twins are virtual models of real objects, places, individuals, or processes. For over 20 years, industries have utilized them to simulate the operation of real-world entities on computers, similar to video games, allowing for performance testing and limitation identification, which fosters advancement across various sectors. In the manufacturing industry, digital twins simulate and optimize production processes, enabling companies to enhance efficiency, reduce costs, and detect potential issues before they arise in reality. Aerospace companies employ digital twins to design and test aircraft prototypes virtually, simulating their operation under different scenarios and conditions, thereby minimizing the time and costs linked to physical prototype development. Similarly, the automotive industry uses digital twins to design vehicles, optimize efficiency, simulate crash tests, and evaluate autonomous driving systems. Beyond these applications, digital twins can model and simulate the behavior of ecosystems, climate change, water resource management, and sustainable urban planning.

The creation and representation of digital twins depend on the application field and the system complexity. In engineering, digital twins use mathematical models with equations that analytically describe the physical system behavior. Translating these equations into computer programs enables their simulation on a computer. In industries like craftsmanship or construction, 3D models and CAD[4] software visually represent digital twins and simulate their functionality in various contexts. Combining approaches – equations, computer programs, 3D models, and CAD – results in more advanced digital twins.

AI is not essential for creating digital twins but greatly enhances their development. Theory-driven AI models help build moderately complex digital twins. Digital twins often use sensor data from the physical system, with machine learning helping to collect and analyze that data. Analyzing this data deepens understanding of the physical system, allowing the creation of more complex and accurate digital twins. These models can then perform realistic simulations to predict system behavior in various scenarios. Machine learning also identifies which parameters in the digital twin equations mainly influence its behavior (feature importance, see Chapter 3, Section 3.5). Furthermore, integrating learning techniques from machine learning into digital twins allows the exploration of learning processes in biological systems. While biological systems and AI learn differently – for example, AI processes many examples to recognize patterns, whereas the human brain learns with fewer examples – both face common challenges in balancing supervised, unsupervised, and reinforcement learning (see Chapter 1, Section 1.1). Comparative studies between these systems reveal strategies the brain uses to solve these challenges [Caligiore et al. 2019].

For some years, the concept of digital twins has gained traction in healthcare, driving projects in pharmaceutical development and digital replication of human anatomy. For example, Dassault Systèmes[5] 3DExperience Lab has developed platforms to represent anatomical structures digitally. In medicine, digital twins integrate a patient's physiological and behavioral data through data-driven and theory-driven approaches. Soon, digital twins will allow researchers to simulate disease effects on computers without human intervention, enabling *personalized drug* development that targets the patient's specific condition while minimizing standard treatment side effects. Combining AI insights with advanced 3D bioprinting also creates possibilities for regenerating damaged organs, enhancing patients' quality and longevity. Using a digital twin, researchers can predict drug impacts, physicians can simulate surgeries or test diagnostic procedures, and patients can understand how lifestyle changes may affect their health.

Digital twins will not act as sentient beings capable of emotions but instead will serve as functional models of the *brain-body-environment system*. These models simulate how health might respond to specific lifestyle or environmental changes. Particularly promising, digital twins allow the rapid, ethical, and cost-effective evaluation of new personalized therapies. For example, determining the ideal drug dosage for a Parkinson's patient – balancing minimized side effects with reduced symptoms like tremors – becomes more efficient by testing various doses on a digital twin rather than on the patient, which avoids discomfort, high costs, and delays. The dose producing the best result in the digital twin likely matches the optimal one for the patient, enabling highly tailored therapies based on the patient's unique digital profile (see Figure 2.2). More broadly, digital

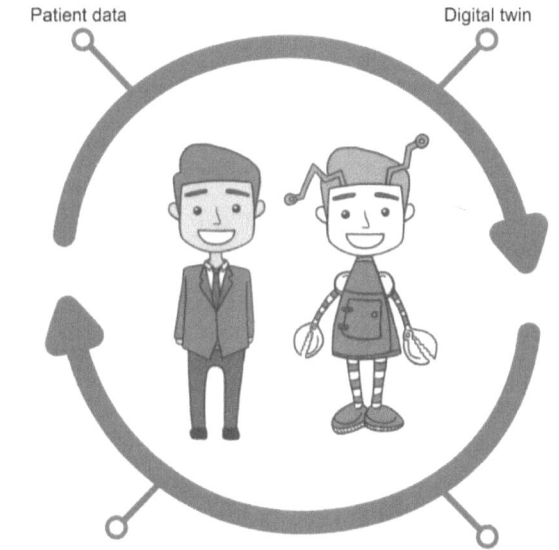

FIGURE 2.2   Utilizing the digital twin to select the optimal therapy. A model is created based on patient data (clinical, anatomical-physiological, biographical, etc.) that reflects the patient's brain and body functioning (digital twin). Various therapies are tested on the digital twin to identify the best-customized treatment for the patient.

Source: Caligiore [2022].

twins also support basic research into other living beings, potentially allowing direct testing of new drugs on their digital counterparts, thus significantly reducing, and perhaps eventually eliminating, the need for animal testing (see Chapter 4, Section 4.3).

One approach to developing digital twins involves using theory-driven AI alongside a technique known as "computational phenotyping" [Schwartenbeck and Friston 2016]. For example, an experiment could aim to study the motor behavior of individuals with Parkinson's disease. Researchers might design an eye-hand coordination task where participants view a sequence of sample movements that they must replicate. Below are the steps to apply the computational phenotyping method:

1. Data collection from real subjects: Collect data on the behavior of participating patients, focusing on performance metrics such as fluency and speed of movement execution related to the specific motor tasks assigned. Additionally, researchers can gather data on the subjects' neural activity. This second type of data can enhance the effectiveness of computational phenotyping, but it is not essential for applying the method.

2. Writing digital twin equations: This process involves building a theory-driven AI system through a computational model that emulates the interactions among the brain regions most relevant to the task. For instance, when participants engage in an eye-hand coordination task, the computational model must mathematically describe how specific brain areas involved in motor control and vision interact. Constructing this model requires a set of equations, such as differential equations, whose solutions indicate how the activity of a particular brain region varies over time in response to the other regions activities. The values of some model parameters derive from previously documented data. For example, scientific literature indicates that neurons in a specific brain area take 20 milliseconds to deactivate after stimulation, providing a parameter for the model equations. However, other parameters, referred to as "free parameters", must be determined using static techniques. Developers can implement the model through software, allowing for examination of its functioning via computer simulations.

3. Finding the free parameters of the model: Statistical methods approximate the values of free parameters to align the model behavior with that of subjects participating in the experiment (*data fitting*). For instance, statistical techniques may indicate that to replicate the performance of a subject in a hand-eye coordination task, the model-free parameters p1 and p2 should be set to 0.2 and 0.9, respectively. These parameters have biological significance and could represent the concentration levels of two neuromodulators. Determining these values customizes the computational model for the specific subject, transforming it into a digital twin capable of emulating the same behavior observed in the subject. Given their biological relevance, the model parameters could allow us to make predictions on the neurophysiological processes in the subject that drive this behavior.

For each participant in the experiment, a set of estimated free parameters will be produced, each corresponding to a specific biological factor. For example, one parameter may represent muscular force, while another may indicate nerve conduction velocity. Different digital twins correspond to different estimated free parameters associated with each real subject. Analyzing these parameters reveals the complex neurocomputational processes underlying the behavior studied in the experiment. For instance, the model might suggest that a particular deficit in hand-eye coordination relates to an alteration in a specific neuromuscular parameter. Thus, the digital twin represents a highly efficient and cost-effective tool for formulating detailed hypotheses about the neural processes driving certain behaviors. Using data-driven machine learning approaches, researchers can explore new correlations between the parameters of the theory-driven model representing the digital twin or identify which parameters are most critical to its functioning [Huys, Maia and Frank 2016]. For example, machine learning analysis could reveal a positive correlation between the parameter representing dopamine and the nerve conduction velocity parameter or indicate that nerve conduction velocity is more influential than muscle strength in achieving good hand-eye coordination.

In addition to studying the neurophysiological mechanisms underlying diseases and simulating the effects of potential therapies, digital twins can create detailed simulations of complex medical procedures. These simulations are valuable for medical training, allowing practitioners to gain experience in a virtual environment before encountering real-life situations (see Chapter 4, Section 4). Digital twins can also represent molecules, proteins, or biological processes to accelerate drug discovery and development, reducing time and costs. For instance, digital twins played a crucial role during the coronavirus (COVID-19) pandemic. The synergy between AI and bioinformatics (the science of developing algorithms and software tools for analyzing biological data) significantly shortened the time needed to find a vaccine. While these technologies have not yet eliminated the clinical trial phases, AI-generated predictions of the virus structure saved researchers months of work, reducing the time required to develop a prototype vaccine for human testing. For example, predicting the secondary structure of the RNA sequence of SARS-CoV-2 (the virus responsible for COVID-19) took only 27 seconds instead of the traditional 55 minutes [Arshadi et al. 2020; Jakhar and Kaur 2020]. Moreover, these technologies enabled fluid and air particle spread simulations, assessing the transmission risks in various environments, including schools, offices, theaters, and hospitals. They have also been used to predict disease trends, expedite genome sequencing, and facilitate faster diagnoses with AI-based software capable of analyzing hundreds of scans to detect coronavirus in as little as 20–30 seconds, compared to the traditional 10–15 minutes without AI [Keshavarzi Arshadi et al. 2020].

The concept of the digital twin marks a significant milestone in the evolution of digital technologies. Creating a digital replica of physical objects, industrial processes, or even biological systems has opened up new possibilities across various fields, from manufacturing to medicine. Digital twins enable the monitoring, simulation, and optimization of physical realities in unprecedented ways. However, digital evolution continues, and in recent years, a new paradigm called the 'metaverse' has started to take shape.

While the digital twin focuses on replicating specific objects or processes, the metaverse represents a broader concept aimed at creating shared virtual worlds often populated by avatars, where social and cultural experiences can take place. These digital realms are beginning to blur the lines between physical and virtual reality, paving the way for new forms of interaction and collaboration. The term "metaverse" combines the Greek word "meta", meaning "beyond", with "universe", literally translating to "beyond the universe" – a virtual realm that extends beyond our physical world. This term describes virtual and digital spaces accessible via the Internet, where people can interact. It resembles a vast online city where individuals can create characters, or avatars, that represent them and engage in several activities. In this virtual environment, users can meet others, chat, play games, work, shop, and even create their own objects.

The metaverse is founded on the convergence of three key technologies: Virtual reality, which creates entirely digital environments; augmented reality, which overlays digital elements onto the real world; and AI. Together, these technologies have the potential to revolutionize the traditional healthcare paradigm, paving the way for innovative approaches to care while significantly reducing healthcare costs. While the digital twin emphasizes a detailed and functional replication of the brain-body-environment system, the metaverse actively engages the patient, enhancing their perception of reality through an immersive experience. For instance, consider the application of the metaverse in postoperative rehabilitation. Instead of adhering to conventional rehabilitation programs in a physical clinic, patients could use virtual reality devices within the metaverse to perform rehabilitation exercises from home, immersing themselves in a virtual environment. Therapists could also join the same environment to monitor progress and make real-time adjustments, providing highly personalized care. Numerous scientific articles and research projects have demonstrated that immersing patients in a virtual world can significantly alleviate pain and assist in managing various anxiety and stress-related situations, including postoperative recovery [Dy et al. 2023; Spiegel et al. 2019]. Another application of the metaverse in healthcare pertains to training. Medical students could use it to simulate complex surgical procedures in a virtual setting, allowing them to acquire practical skills safely before confronting real-life scenarios. This approach would not only enhance patient safety but also accelerate the learning process for future doctors and improve the overall quality of medical training.

Whereas virtual reality emphasizes individual immersion in digital environments, the metaverse encompasses a *shared virtual world* where users can interact as in real life. This *social aspect* is a defining characteristic of the metaverse, which can be applied in various ways within the medical field. Another key feature of the metaverse is the *fusion of physical and virtual worlds*: Actions in the real world affect experiences in the virtual world, and vice versa [Park and Kim 2022; Riva and Wiederhold 2022]. For instance, if a movement is made in the real world, the virtual avatar replicates that movement. Conversely, interacting with the avatar in virtual space generates a tactile response in the physical body. This interconnectedness between the physical and virtual realms is perhaps the most crucial component of the healthcare metaverse. To link real physical environments with digital ones, metaverse applications can harness the synergy of multiple technologies, including AI, blockchain,[6] and the Internet of

Things (IoT),[7] facilitating seamless interactions between virtual and real-world experiences. In therapeutic settings, virtual reality has been used to create the illusion of an object presence even when it is absent. By integrating various embedded technologies – such as haptic and interoceptive technologies that detect internal biological signals like heartbeat and body temperature – with enhanced visualization capabilities from first- or third-person perspectives and big data analysis through machine learning algorithms, the metaverse offers a more profound immersive experience than virtual reality alone [Huynh-The et al. 2023].

The hybrid nature of the metaverse, blending real and virtual elements, intensifies a profound sense of presence – more so than virtual reality alone. This experience "tricks" our senses and engages the brain "predictive coding systems", which govern our perception of the body. Predictive coding, a process identified in neuroscience, allows the brain to anticipate sensory events before they occur [Clark 2014]. Similarly, the metaverse creates a hybrid environment that enables users to interact and explore the sensation of being present. In simpler terms, the metaverse prompts our brains to anticipate sensory reactions to the actions of other users in the digital realm, mimicking the responses we expect in the physical world [Riva and Wiederhold 2022]. Predictive coding suggests that the brain actively generates an internal model of the body and its environment; when this simulation malfunctions, it can lead to various mental health issues [Riva et al. 2021]. By replicating these mechanisms, the metaverse fosters a strong sense of ownership over a digital body (or its parts), which can be crucial for several medical applications.

Let us discuss some examples that show how blending physical and virtual worlds, combined with social interaction, could make the metaverse a powerful tool in healthcare. Wearable devices and medical sensors can be linked to avatars in the metaverse, enabling continuous monitoring of patients. Wearable devices and medical sensors can be connected to avatars in the metaverse to monitor patients conditions. The collected data, such as blood pressure, heart rate, and blood sugar levels, can be visualized in real time within the virtual world, enabling both patients and doctors to monitor these vital signs. For patients undergoing physical therapy or rehabilitation, the metaverse offers the opportunity to participate in virtual therapy sessions. A patient recovering from surgery, for example, can use virtual reality devices to perform interactive rehabilitation exercises, with their progress monitored and shared with therapists in the physical world. Patients with chronic medical conditions can also join virtual support groups within the metaverse. These groups provide a safe, interactive space for patients to share experiences, exchange advice, and receive emotional support from others who are facing similar challenges, as well as from healthcare professionals. For example, patients can participate in group therapy virtual sessions where they interact with therapists and fellow participants through avatars. This approach may offer a more accessible and comfortable experience than face-to-face interactions. Consider a group of individuals with social anxiety who wish to engage in group therapy: Many would likely struggle to participate in person due to their condition. In this scenario, group therapy in the metaverse could provide a viable solution. Each participant creates an avatar to represent their virtual identity, with the ability to customize the avatar in various ways, allowing for personal expression. The therapy session takes place in a secure and comfortable virtual environment designed to resemble a therapy room where participants and the therapist, represented by their avatars, can gather. Participants communicate through their avatars, either via voice or text chat. In this way, they could share experiences, express emotions, and discuss challenges related to social anxiety. The avatars help reduce the feeling of being judged or observed, lowering anxiety typically associated with in-person interactions. This approach can encourage greater openness and sharing within group discussions. Patients dealing with body dysmorphia (e.g., eating disorders) or social difficulties (e.g., autism) could particularly benefit from this technology [Cerasa et al. 2022]. Since therapy takes place in a digital environment, patients can join sessions from the comfort of their own homes or any other location they prefer, eliminating the need for physical travel, which can be a barrier for some individuals. The therapist, represented by an avatar, guides the session, offers emotional support, provides tools for managing social anxiety, and facilitates group discussions. This example demonstrates how the metaverse can give an accessible and comfortable alternative to traditional group therapy, allowing individuals with social anxiety to receive therapeutic support without the overwhelming pressure of physical presence.

The social aspect of the metaverse opens new opportunities for medical training and collaboration. For example, medical students and healthcare professionals can use the metaverse to engage in collaborative training sessions, simulate complex medical scenarios, participate in clinical case discussions, and learn from peers in a shared virtual environment. Medical conferences can also be hosted in the metaverse, allowing physicians and researchers to attend virtual events. They could virtually listen to speakers, network with colleagues, and explore virtual exhibitions showcasing the latest medical technologies and research. Scholars from around the world can collaborate in shared virtual spaces in the metaverse to conduct joint medical research, share data, and discuss breakthroughs. Additionally, healthcare professionals can perform realistic medical emergency simulations in the metaverse, enabling multidisciplinary teams to practice handling critical situations and improve their emergency response skills. The metaverse also provides telemedicine opportunities for patients in remote areas or those who find it difficult to visit a medical facility. They can access high-quality healthcare through virtual consultations, avoiding long-distance travel. Moreover, the metaverse can create immersive educational experiences for patients. For example, patients can virtually explore their bodies and simulate medical conditions, offering them a deeper understanding of their illnesses and treatment options. This virtual environment provides a level of detachment, helping reduce the anxiety that might come from observing the effects of a disease in real life. Patients with rare autoimmune diseases could participate in virtual simulations that allow them to visualize how the disease affects their body (represented by an avatar). This immersive experience can help patients better understand their condition and communicate more effectively with healthcare providers about their needs and treatment options.

The potential applications of digital twins and the metaverse discussed in this section highlight their crucial role in enhancing patients self-awareness. These tools could help improve understanding of how our bodies function, the nature of medical conditions, and available treatment options. Alongside digital twins and the metaverse, real-life models of miniature organs, known as "organoids", can also simulate body functions. Organoids are three-dimensional, self-organizing structures grown in vitro from pluripotent[8] stem cells or differentiated[9] adult cells. These systems replicate the functional and structural characteristics of the tissues or organs they represent. The first organoids were derived from intestinal stem cells, forming structures similar to the human intestine. Since then, organoids for various systems – such as the liver, kidneys, lungs, and heart – have been developed. Recently, researchers have begun developing brain organoids, or "mini-brains", to study brain function and neurological diseases. Brain organoids can exhibit some neuronal features, like the formation of synapses and the production of neuromodulators. The term "organoid intelligence" refers to their ability to show limited cognitive or neuronal activity, but the concept of fully functional artificial brains remains a distant hypothesis. These organoids are still too simple compared to human brains. They have limited neuronal activity and raise significant ethical concerns, such as the risk of creating artificial life and the debate over whether they could be considered conscious entities (see Chapter 3). It is also possible to create "organoid systems" or "multi-organ organoids" composed of different types of organoids that communicate with each other. These systems could replicate the interaction between organs, as in the human body. In this way, it could be possible to study how diseases spread between organs and lead to the development of new therapies targeting multiple organ systems.

AI can enhance the creation of organoids [Badai, Bu and Zhang 2020]. AI can analyze data from imaging techniques and cell and molecular biology to support the modeling of cell growth and optimize culture conditions. For example, AI can identify the most promising cells for organoid generation using feature importance techniques (see Chapter 3, Section 3.5). Although digital twins and organoids are still in development and require further research to understand their potential and limitations, these technologies have opened new avenues for research. They allow for faster, more personalized, and more efficient studies of organ development, disease modeling, and the analysis of new therapies.

2.3 AI FOR EARLY DIAGNOSIS

A recent article in the prestigious journal *Nature Medicine* discusses the potential of AI to assist physicians in diagnosing common childhood illnesses, ranging from the flu to meningitis, by analyzing patients symptoms, medical history, lab results, and other clinical data [Liang et al. 2019]. In recent years, numerous scientific publications have proposed machine learning algorithms that, by processing large datasets, can help doctors identify new markers for the early and accurate diagnosis of various diseases [Myszczynska et al. 2020]. To demonstrate this capability, we will explore some applications of machine learning in diagnosing Alzheimer's disease, the most prevalent neurodegenerative disorder worldwide. Alzheimer's primarily affects the central nervous system, impairing neural mechanisms critical for cognitive function. Symptoms often begin with occasional short-term memory loss – such as forgetting the name of someone you have just met or a recently made shopping list – but gradually extend to long-term memory, which stores events and skills learned over extended periods. As the disease progresses, patients may experience disorientation and difficulties with communication and reasoning.

They may also experience difficulty performing daily tasks and behavioral changes like irritability and apathy. On a neural level, Alzheimer's patients show abnormal protein accumulations in the brain, notably beta-amyloid plaques and tau tangles. These protein deposits damage nerve cells, disrupt synaptic connections, and lead to neuronal death. Consequently, progressive atrophy occurs, especially in brain regions responsible for memory and higher cognitive functions.

The exact triggers of Alzheimer's disease remain unclear, but the *cholinergic hypothesis* points to reduced levels of acetylcholine – a neurotransmitter vital for nerve cell communication – as a critical factor [Hardy 2006]. Acetylcholine is crucial for memory, learning, and overall central nervous system function. Its decline has been linked to the formation of beta-amyloid plaques and the accumulation of tau tangles [Mesulam 2004]. *Dopamine*, a neurotransmitter crucial for movement control (as discussed in the context of Parkinson's disease), also plays an important role in regulating mood and cognitive functions. It may contribute to the progression of Alzheimer's as well [Caligiore et al. 2020; Nobili et al. 2017]. Recent research has uncovered how conditions seemingly unrelated to cognition could serve as risk factors for Alzheimer's disease. For example, substantial evidence suggests that mechanisms related to oral hygiene [Pruntel et al. 2024] and *gut health*, including dysbiosis, may influence the disease pathogenesis [Pistollato et al. 2016]. In particular, the interaction between gut microbiota and the central nervous system may significantly contribute to the development of Alzheimer's [Caligiore, Giocondo and Silvetti 2022]. Increasingly, Alzheimer's is seen as a *systemic disease* in which various neurochemical and physiological factors interact in complex and poorly understood ways, promoting the formation of characteristic lesions and the progression of symptoms.

Current diagnostic techniques for Alzheimer's disease often fail to account for its systemic nature. Physicians typically base diagnoses on medical history, clinical and neurological exams, and brain activity analysis. These methods could involve invasive procedures like cerebrospinal fluid analysis or costly brain imaging, requiring highly specialized personnel [Rasmussen and Langerman 2019]. Doctors usually initiate evaluations after the disease has already caused significant brain damage and symptoms are visible. However, neurodegenerative processes and physiological changes can start 10–15 years before clinical signs appear [Amieva et al. 2008; Beason-Held et al. 2013]. This underscores the need for affordable, noninvasive diagnostic methods that can identify individuals at high risk for Alzheimer's long before symptoms emerge. Early diagnosis allows timely intervention, improves disease management, reduces healthcare costs, and enhances treatment outcomes. Providing treatments in the preclinical stage, before symptoms develop, can offer greater benefits. Additionally, early lifestyle changes, such as managing risk factors like hypertension, smoking, obesity, and diabetes, could slow or prevent the disease [Caligiore, Giocondo and Silvetti 2022; Norton et al. 2014]. Early diagnosis also allows individuals to make important decisions about their care, asset management, and plans while they are capable. Families gain time to prepare for caregiving roles[10] and adjust to the patient's behavioral and personality changes. This preparation can reduce stress and lower the risk of psychological issues such as anxiety and depression [De Vugt and Verhey 2013; Frias, Cabrera and Zabalegui 2020].

Recent studies highlight the significant contributions of machine learning to Alzheimer's research and clinical practice, enabling highly reliable predictions through personalized data [Grassi et al. 2019; Merone et al. 2022; Moustafa 2021]. AI algorithms designed to identify novel markers for the early diagnosis of Alzheimer's disease utilize a wide array of data from diverse sources, including genetic information, cerebrospinal fluid biomarkers, brain imaging, demographic and clinical data, and cognitive assessments [Dukart, Sambataro and Bertolino 2016; Grueso and Viejo-Sobera 2021; Liu et al. 2018; Long et al. 2017; Platero, Lin and Tobar 2019]. However, some data used to train these algorithms – such as cerebrospinal fluid biomarkers – are costly and can only be obtained through invasive procedures. As a result, while machine learning research for early Alzheimer's diagnosis based on such data is valuable as a proof of concept, it cannot yet replace conventional techniques, as it shares similar complexities and costs.

To address current limitations, recent studies have proposed machine learning algorithms that rely solely on *noninvasive, easily collected data*, such as neuropsychological test scores, sociodemographic information, clinical data, and blood biomarkers [Grassi et al. 2019; Merone et al. 2022]. A team of American researchers recently developed an AI capable of analyzing natural language to detect subtle signs of cognitive decline. This simple AI-based language test can accurately predict which cognitively normal individuals may develop Alzheimer's disease [Eyigoz et al. 2020]. In a similar vein, a collaboration between the Institute of Cognitive Sciences and Technologies (ISTC) of the Italian National Research Council (CNR), the IRCCS Fondazione Santa Lucia, the University Campus Bio-Medico, and the Sapienza University of Rome led to the development of a machine learning algorithm that can predict Alzheimer's onset up to nine years before clinical symptoms appear. This algorithm relies on inexpensive data from neuropsychological tests [Merone et al. 2022]. The algorithm was trained using 69 tests from over 500 individuals (some healthy, others likely to develop Alzheimer's within nine years), with data drawn from the public ADNI[11] database. The tests covered reading skills, semantic word processing, memory effects of dementia, problem-solving, self-care, and mood assessment, along with indirect measures of neurological, cardiovascular, respiratory, gastrointestinal, renal, and cerebellar function, which plays a key role in movement and emotion regulation [Paulin 1993; Schmahmann and Caplan 2006]. The AI determined that only five of the 69 tests were crucial for predicting the disorder onset within nine years. Two of these tests are already standard in Alzheimer's diagnosis: The ADAS (Alzheimer's Disease Assessment Scale), which measures cognitive impairment, and the CDR (Clinical Dementia Rating), which evaluates dementia severity based on memory, orientation, problem-solving, and social activities. The other three tests identified by the AI focus on medical history, particularly head, eye, ear, nose, throat, and kidney problems, as well as cerebellar dysfunction. This discovery is surprising, as it suggests that seemingly unrelated conditions – such as head injuries and kidney or cerebellar issues – may be critical disease predictors. These findings highlight the need to view Alzheimer's as a systemic disease, emphasizing the interactions between the brain and the rest of the body.

Several studies indirectly support the findings of the machine learning model. Head injuries, for instance, may have long-term effects on cognitive function, with research suggesting that older adults who have suffered head injuries experience faster cognitive decline than those who have not [Luukinen et al. 1999; Whiteneck, Gerhart and Cusick 2004]. Traumatic brain injuries, especially those occurring in early or middle age, may also increase the risk of developing Alzheimer's later in life [Plassman et al. 2000]. Similarly, renal dysfunction has been linked to memory impairment and accelerated cognitive decline [Buchman et al. 2009; Etgen 2015]. There is also evidence connecting cerebellar dysfunction with early-onset Alzheimer's [Jacobs et al. 2018; Testi et al. 2014]. While these studies provide empirical support for the machine learning model results, further research on humans is needed to directly confirm its findings before it can be applied in healthcare. Nevertheless, the model represents an important first step toward developing a preventive screening tool based on machine learning, which could be incorporated into routine health checkups, providing a fast and effective method for early identification of potential health risks.

In a similar study using a theory-driven approach, researchers from ISTC-CNR, IRCCS Fondazione Santa Lucia, and the University Campus Bio-Medico in Rome demonstrated how an AI model can simulate the neural mechanisms underlying the early stages of Alzheimer's disease [Caligiore et al. 2020]. The model revealed that dysfunction in the ventral tegmental area (VTA) – a deep brain region primarily composed of dopamine-producing neurons – could be one of the earliest events linked to Alzheimer's. It also highlighted the importance of studying the interaction between dopamine and noradrenaline, a neuromodulator crucial for memory and learning, produced by neurons in the locus coeruleus (LC). The AI simulations showed that early neurodegeneration in the VTA, and the resulting decline in dopamine release, might not immediately lead to noticeable behavioral symptoms. This is because an initial increase in LC activity, and consequently in noradrenaline release, compensates for the reduced dopamine. However, this *compensatory mechanism* is short-lived. As VTA degeneration worsens, LC activity also declines, at which point the simulated subjects begin to display early symptoms of Alzheimer's, such as memory loss. The model suggests that monitoring these compensatory mechanisms in real individuals – starting from the ages of 50–60, before the typical onset of Alzheimer's symptoms around age 65 – could aid in extremely early diagnosis. Since VTA activity is also linked to emotional regulation and motivation, the AI model underscores the role of psychological factors. It suggests that reduced motivation and loss of interest, often overlooked by patients and families, may accelerate Alzheimer's progression. This research, published in the *Journal of Alzheimer's Disease* [Caligiore et al. 2020], opens new possibilities for early diagnosis and the development of therapies targeting dopamine-related brain regions in the disease's early phases, potentially slowing or halting neurodegeneration.

Overall, these findings highlight the importance of studying Alzheimer's disease from a systems-level perspective, focusing on the interactions between brain and body dysfunctions as critical factors for early diagnosis. Researchers can extend this approach to other diseases as well. Machine learning can help identify new markers for the early detection of various conditions, including cancer [Farinella et al. 2022; Jones et al. 2022], diabetes [Rawat et al. 2022], and many others [Ahsan, Luna and Siddique 2022]. This innovative technology has the potential to revolutionize diagnostics, improving treatment outcomes and recovery chances for patients.

2.4 AI FOR COMPUTER-BASED TESTING OF NEW PERSONALIZED THERAPIES

AI can facilitate testing new personalized and systemic therapies – treatments tailored to the specific individual while considering the brain-body-environment system. This approach offers numerous opportunities to accelerate and improve the effectiveness of medical treatments. For example, AI can analyze large genomic datasets to identify patterns and genetic mutations linked to diseases. This insight can guide the development of targeted therapies designed to treat a patient's unique genetic profile. Consider a database containing the genetic information of individuals with breast cancer. This dataset might include DNA from patients with different subtypes, such as estrogen receptor-positive[12] (ER+), human epidermal growth factor receptor 2-positive (HER2+), and triple-negative breast cancer (TNBC). AI could be trained with unsupervised algorithms (see Chapter 1, Section 1.1) to detect genetic mutations associated with each breast cancer subtype. Once trained, the AI algorithm could group the data into distinct clusters, allowing researchers to identify meaningful patterns. For instance, a cluster might predominantly consist of patients with ER+ tumors who exhibit overexpression[13] of a specific gene. Now imagine a newly diagnosed patient with ER+ breast cancer. Doctors could take a sample of her tumor tissue and sequence its DNA. AI, trained to recognize genetic mutations associated with different breast cancer subtypes, would then analyze the sequencing results to identify the specific mutations in the patient tumor. Through this process, doctors might discover that the patient's tumor has a mutation linked to a positive response to a targeted drug known as an 'ER receptor inhibitor'. This drug is suitable for patients whose genetic mutations lead to excessive activation of the ER receptor. Based on this finding, doctors could prescribe a personalized treatment that includes the ER receptor inhibitor. With this targeted therapy, tailored to the unique genetic profile of her tumor, the patient is more likely to experience positive treatment outcomes with fewer side effects compared to generic treatments. In this example, AI-driven analysis of genomic data enabled physicians to make informed decisions about the patient's treatment, offering her a customized and optimized therapy based on the specific genetic characteristics of her breast cancer.

AI enables the simulation of new systemic therapies through digital twins. These virtual models allow doctors to assess the effectiveness of different treatments, reducing the risks and costs associated with testing directly on patients. A notable example of this approach is a study we conducted at ISTC-CNR involving the students from the Advanced School in AI.[14] The findings, published in the international journal *Frontiers in Systems Neuroscience*, aimed to develop a theory-driven model to study different types of tremor in Parkinson's disease, ultimately suggesting new therapeutic approaches [Caligiore et al. 2021]. Tremor is one of the main symptoms of Parkinson's disease, characterized by involuntary oscil-latory movements, typically starting in one limb (often a hand or foot) and spreading to other parts of the body. Tremor is linked to a dopamine deficiency in the substantia nigra, a brain region responsible for move-ment control. Dopamine is critical in regulating movement and main-taining balance between the brain motor control systems. In Parkinson's disease, dopamine-producing cells in the substantia nigra degenerate and die, leading to a deficiency that disrupts movement control. This process causes tremor and other motor issues, such as difficulty initiating move-ment, slowness, and problems with balance and posture. Many current Parkinson's treatments focus on restoring dopamine levels. While these drugs are effective for most motor dysfunctions, their impact on tremor varies: In some patients, it is not reduced despite treatment.

Recent studies suggest that different types of tremors, linked to distinct pathophysiological mechanisms across a network of brain regions, may explain the lack of treatment efficacy in these patients. These areas include regions beyond those traditionally associated with dopamine release and not directly related to the disease. For instance, in addition to dopamine deficiency, deficits may occur in the cerebellum and thalamus [Caligiore et al. 2016; Dirkx et al. 2016; Wu and Hallett 2013], or there may be dys-functions in the production of other substances like serotonin [Caligiore, Giocondo and Silvetti 2022; Wilson et al. 2019]. Serotonin is a neuromodu-lator that is important for regulating mood and movement. There is a close interplay between dopamine and serotonin in the brain, with serotonin modulating dopamine release [Benloucif, Keegan and Galloway 1993]. Low serotonin levels can influence the dopaminergic system, contribut-ing to various motor symptoms, including tremor. Therefore, tremor in Parkinson's disease can arise from multiple factors, including dopamine deficiency and reduced serotonin levels. However, as dopamine and sero-tonin are intricately interconnected, a deficiency in one neurotransmitter can affect the other, leading to different forms of tremor.

The bioinspired computational model developed at ISTC-CNR explores the neural mechanisms underlying a potential type of tremor primarily involving serotonin and dopamine binding [Caligiore et al. 2021]. The model simulates the interactions between different brain regions associated with Parkinsonian tremor (system-level approach) and generates signals to control the movement of a computer-simulated anthropomorphic robotic arm. Simulations show that a physiological increase in serotonin can partially restore dopamine levels during the early stages of the disease, before the onset of full-blown tremor. This result suggests that monitoring *serotonin* concentration could be critical for *early diagnosis*. The simulations also indicate the potential effectiveness of a novel pharmacological treatment that targets serotonin to restore dopamine levels (see Figure 2.3). While this result has been validated by reproducing data from other studies on human patients, the model findings on the potential for serotonin-based therapy are only an initial step, intended to encourage further pharmacological research. Additional well-established procedures, including further analyses and clinical trials (see following pages), will be necessary before testing new drugs. This example illustrates the importance of computational models that simulate the interactions of a network of brain areas rather than focusing on a single area traditionally studied for a specific disease. Such models are essential for exploring new therapeutic strategies. This system-level approach could be applied to study healthy individuals or to investigate other diseases.

Machine learning models (data-driven approaches) can be crucial in the pharmaceutical field. Many research laboratories, pharmaceutical companies, and biotechnology firms are actively exploring using AI to develop software that predicts the pharmaceutical properties of new molecular compounds, a process known as *drug discovery*. The development of pharmaceutical products involves clinical trials, which are notoriously lengthy and costly. Clinical trials are scientific studies conducted on human participants to evaluate the safety and efficacy of new drugs or medical treatments. These trials are divided into several phases, beginning with laboratory experimentation, followed by testing on a small group to assess safety and determine the optimal dose. Successively, the drug undergoes testing on progressively greater groups of participants. Clinical trials are essential to ensure that drugs are effective and safe before being approved for public use. However, the process is time-consuming and expensive due to the rigorous testing and evaluation required. Integrating AI into this process could transform the industry by making drug discovery more affordable, faster, and safer, benefiting patients and healthcare providers. The combined use of AI and traditional clinical trials could enable researchers to efficiently identify new compounds with pharmaceutical

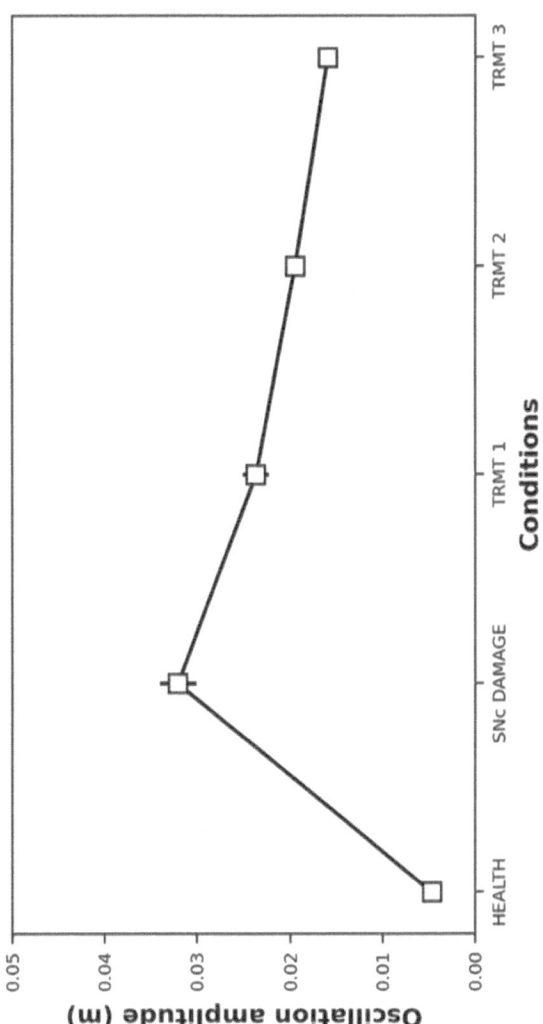

FIGURE 2.3 Reduction in tremor in the computational model proposed by Caligiore and colleagues [2021] following the application of a treatment that targets serotonin rather than dopamine. The x-axis represents five conditions: A healthy model (Healthy), a model with a dopamine deficit in the substantia nigra (SNc) that induces tremor (Parkinson's), and three models simulating the effects of increasing doses of a serotonin-targeting pharmacological treatment (TRMT 1, 2, 3). Tremor is measured by the oscillations of the robotic arm (y-axis). The figure demonstrates a clear decrease in arm oscillations as the treatment dose increases.

Source: Adapted from Caligiore et al. [2021].

potential (see the example of coronavirus research in Section 2.2). This innovative approach could uncover new applications for previously tested compounds, thereby expanding therapeutic possibilities. Machine learning can accelerate drug discovery by analyzing large chemical compound databases and predicting which combinations are most likely effective against specific diseases [Vamathevan et al. 2019].

2.5 AI FOR DISEASE PREVENTION

AI is also poised to impact disease prevention. As discussed in Chapter 1, AI has been used in medical imaging – such as retinal scans, body surfaces, bones, and internal organs – to support better health maintenance through faster diagnoses and accurate tracking of disease progression. AI applications mentioned earlier, including those for early diagnosis and simulating the effects of innovative therapies, also play a critical role in prevention [Wani et al. 2022]. We have seen how AI can analyze vast datasets to uncover correlations that are difficult to detect with traditional statistical methods. This process aids in the identification of risk factors and disease development patterns. Take, for example, the study of epidemics. Imagine a large dataset of infectious disease cases, such as influenza or COVID-19, containing information on geographical locations, patient ages, symptoms, exposure times, and recurrence rates. Traditional statistical tools might struggle with complexity and nonlinear relationships between these variables. In contrast, an AI-based system could quickly and accurately analyze the data, revealing patterns not easily identified by conventional methods. For instance, it might detect that individuals in an age group who have traveled to specific regions are at a higher risk of developing the disease. Predictive models analyzing epidemic spread can help healthcare professionals make informed decisions about resource allocation, identify high-risk geographical areas, and implement targeted prevention strategies. Without AI, these complex correlations could remain hidden, making disease prevention and epidemic management more challenging.

AI can be more effective than traditional statistical methods in complex situations due to its ability to process vast amounts of data and detect nonlinear correlations. Classical statistics often relies on specific hypotheses and tests to determine whether relationships in the data are statistically significant. This approach works well for testing hypotheses with a small set of predefined parameters. However, traditional statistics assumes that the analyzed data adhere to certain conditions, which can be limiting when dealing with complex or heterogeneous datasets. For example, many statistical tests assume that the data follow a normal (bell-shaped) distribution, that the observations are independent, or that relationships between variables are linear – meaning that changes in one variable are proportional to changes in another. Machine learning develops models that autonomously learn to recognize complex patterns in diverse datasets, often without requiring explicit assumptions about data distribution or independence of observations. However, this aspect does not always make machine learning preferable to traditional statistics; the choice depends on the investigated problem. A good rule of thumb is to choose the most straightforward method that fits the situation. If the data meet the assumptions of traditional statistical methods, they may be more than sufficient. In some cases, combining both approaches can offer deeper insights by highlighting different aspects of the problem.

AI can analyze genomic data to identify genetic predispositions to specific diseases, enabling personalized prevention plans [Juhola et al. 2018; Srinivasu et al. 2022]. For instance, imagine someone undergoing a DNA test for diagnostic purposes. The sequencing of their genome identifies specific genetic variants linked to an increased risk of heart disease and type 2 diabetes. An AI system can then analyze this genomic data, examining the person's genetic variants to assess their future risk of developing these conditions. A machine learning model can be trained as a predictor/regressor (see Chapter 1, Section 1.1) using historical data from individuals with similar genetic profiles who developed heart disease and type 2 diabetes over time. Once trained, the model can be applied to assess the individual risk for the specific person, "X". Based on X genetic variants, the model estimates their likelihood of developing heart disease and type 2 diabetes in the future. Through AI-driven genomic data analysis, it becomes possible to provide personalized risk assessments, allowing for targeted preventive measures – such as lifestyle changes and regular screenings – that can improve long-term health outcomes.

Some companies have developed wearable devices (such as belts, bracelets, etc.) that can collect real-time health data, like heart rate or body temperature. These technologies, often powered by AI, continuously monitor vital signs, such as blood pressure and heart rate, detecting abnormalities and providing early warnings if imbalances are detected. For example, smartwatches and fitness trackers are equipped with sensors like photoplethysmographs to monitor heart rate and accelerometers to track movement. AI algorithms analyze sensor data to detect unusual variations in heart rhythm or other vital parameters. When the device finds abnormalities, it can send real-time notifications to the user or designated individuals (such as family members or healthcare providers), alerting them to a potential issue and facilitating timely intervention if necessary. In this way, AI is critical to analyzing real-time health data, enabling swift responses to potential problems with vital signs.

Monitoring our habits does not always require wristbands or other wearable devices. AI-based virtual assistants can provide recommendations on healthy eating, improving sleep quality, and general support for quitting smoking, losing weight, or managing stress – helping to reduce the risk of chronic diseases. Examples include Google Fit, which uses AI to track physical activity and offer guidance on achieving fitness goals for a healthier lifestyle; Ada, an AI-driven healthcare app that assesses symptoms and provides personalized health advice, identifying risk factors and suggesting lifestyle changes; and Noom, an AI-powered app that helps users lose weight through personalized diet plans and motivational support to achieve health goals. Before using any AI-based health management app, it is important to check its credibility and ensure it has been developed by reputable organizations. Additionally, it is advisable to consult a healthcare professional before making significant lifestyle changes or following health advice from apps.

In this chapter, we explored how AI can serve as a highly innovative tool for the early diagnosis, treatment, and prevention of diseases, ultimately helping to improve people quality of life. However, to fully realize this potential, it is crucial to employ a responsible approach to AI use, ensuring strict user privacy protection and addressing the new ethical and legal challenges that inevitably arise.

NOTES

1. Data from a report by Statista, a company specializing in statistical analysis, can be found at: https://www.statista.com/topics/10011/ai-in-healthcare/#topicOverview.
2. PubMed is a free online service that provides access to an extensive collection of scientific articles in medicine, biology, and related fields. Users can search and explore academic articles, reviews, and research summaries published in scientific and medical journals. PubMed is a widely used resource by healthcare professionals, researchers, and students.
3. Neuromodulators are chemicals in the nervous system that regulate neuronal activity, influencing signal transmission and the strength of connections between neurons (synapses). They are critical in regulating mood, attention, sleep, learning, and other complex brain activities. Neuromodulators can be released by neurons or by other specialized cells within the nervous system.

4. A CAD (Computer-Aided Design) model digitally represents a three-dimensional object. Designers use CAD software to create, visualize, and analyze objects before production, especially in fields like engineering and architecture.

5. For more information, visit the following website: https://www.3ds.com/.

6. "Blockchain" refers to a technology in digital information security and data certification. Blockchain decentralizes recording transactions in a digital ledger across a peer-to-peer network, where computers share files and information directly without central authority. Nodes in the network collaborate to validate and verify each transaction. Cryptographic algorithms protect the security and integrity of the data, preventing unauthorized modifications or falsification.

7. The term "Internet of Things" (IoT) refers to using the Internet not only for human communication but also to allow objects to exchange data about their status and access information from other objects. These smart devices include environmental sensors, medical equipment, home heaters, refrigerators, and more. Through IoT connectivity, objects collect and send data to other devices or data processing systems, receiving instructions or commands from devices.

8. Stem cells are undifferentiated cells, meaning they have not yet developed a specific function but can differentiate into specialized cell types. These cells exist in tissues such as bone marrow, skin, and umbilical cord blood and regenerate damaged or diseased tissue. Pluripotent stem cells can differentiate into many different cell types in the human body. They develop into all three embryonic tissue types: Endoderm (which forms the gastrointestinal tract and related organs), mesoderm (which forms muscle, bone, and blood tissue), and ectoderm (which forms the skin and central nervous system).

9. Differentiated adult cells are specialized cells that arise from stem or precursor cells during embryonic development or within adult tissues. These cells perform specific functions and have well-defined structures based on the tissue they belong to. For example, skeletal muscle cells contract to generate movement, while liver cells produce enzymes and hormones essential for digestion and nutrient metabolism.

10. A caregiver helps another person who may need support due to illness or old age. They assist with daily activities, such as eating or dressing, and may provide emotional support and companionship.

11. Alzheimer's Disease Neuroimaging Initiative (ADNI) (http://adni.loni.usc.edu).

12. Receptors act like antennae on the surfaces of our body cells, detecting chemical signals from other cells. When receptors recognize the correct chemical signal, they trigger various actions within the cell. For instance, they can prompt a cell to divide, release specific chemicals, or carry out other activities essential for body functions.

13. A gene is "overexpressed" when it produces more protein than usual. Each gene in our body provides instructions to make a specific protein, and the production of this protein is tightly regulated. However, genetic mutations

or other factors can cause some genes to produce excessive proteins. Gene overexpression can affect cell behavior, leading to uncontrolled cell growth or other changes that may contribute to disease development, including certain types of cancer.

14. The Advanced School in Artificial Intelligence (AS-AI) (www.as-ai.org) is a postgraduate institution focused on the study and interdisciplinary application of Artificial Intelligence. AS-AI is organized and supported by AI2Life srl (www.ai2life.com) and the ISTC-CNR (www.istc.cnr.it).

REFERENCES

Ahsan, M. M., Luna, S. A. & Siddique, Z. [2022], *Machine-Learning-Based Disease Diagnosis: A Comprehensive Review*, Healthcare, 10, n. 3, p. 541.

Amieva, H., Le Goff, M., Millet, X., Orgogozo, J. M., Pérès, K., Barberger-Gateau, P., … & Dartigues, J. F. [2008], *Prodromal Alzheimer's Disease: Successive Emergence of the Clinical Symptoms*, Annals of Neurology, 64, n. 5, pp. 492–498.

Arshadi, A. K., Webb, J., Salem, M., Cruz, E., Calad-Thomson, S., Ghadirian, N., … & Yuan, J. S. [2020], *Artificial Intelligence for COVID-19 Drug Discovery and Vaccine Development*, Frontiers in Artificial Intelligence, 3, 65, doi: https://doi.org/10.3389/frai.2020.00065.

Badai, J., Bu, Q. & Zhang, L. [2020], *Review of Artificial Intelligence Applications and Algorithms for Brain Organoid Research*, Interdisciplinary Sciences: Computational Life Sciences, 12, pp. 383–394.

Beason-Held, L. L., Goh, J. O., An, Y., Kraut, M. A., O'Brien, R. J., Ferrucci, L. & Resnick, S. M. [2013], *Changes in Brain Function Occur Years Before the Onset of Cognitive Impairment*, Journal of Neuroscience, 33, n. 46, pp. 18008–18014.

Benloucif, S., Keegan, M. J. & Galloway, M. P. [1993], *Serotonin-Facilitated Dopamine Release In Vivo: Pharmacological Characterization*, Journal of Pharmacology and Experimental Therapeutics, 265, n. 1, pp. 373–377.

Buchman, A. S., Tanne, D., Boyle, P. A., Shah, R. C., Leurgans, S. E. & Bennett, D. A. [2009], *Kidney Function is Associated with the Rate of Cognitive Decline in the Elderly*, Neurology, 73, n. 12, pp. 920–927.

Caligiore, D. [2022], *IA istruzioni per l'uso*, Bologna, Il Mulino.

Caligiore, D., Arbib, M. A., Miall, R. C. & Baldassarre, G. [2019], *The Super-Learning Hypothesis: Integrating Learning Processes Across Cortex, Cerebellum and Basal Ganglia*, Neuroscience & Biobehavioral Reviews, 100, pp. 19–34.

Caligiore, D., Giocondo, F. & Silvetti, M. [2022], *The Neurodegenerative Elderly Syndrome (NES) Hypothesis: Alzheimer and Parkinson Are Two Faces of the Same Disease*, IBRO Neuroscience Reports, 13, pp. 330–343.

Caligiore, D., Helmich, R. C., Hallett, M., Moustafa, A. A., Timmermann, L., Toni, I. & Baldassarre, G. [2016], *Parkinson's Disease as a System-Level Disorder*, NPJ Parkinson's Disease, 2, n. 1, pp. 1–9.

Caligiore, D., Mannella, F., Arbib, M. A. & Baldassarre, G. [2017a], *Dysfunctions of the Basal Ganglia-Cerebellar-Thalamo-Cortical System Produce Motor Tics in Tourette Syndrome*, Plos Computational Biology, 13, n. 3, p. e1005395.

Caligiore, D., Montedori, F., Buscaglione, S. & Capirchio, A. [2021], *Increasing Serotonin to Reduce Parkinsonian Tremor*, Frontiers in Systems Neuroscience, 15, p. 682990.

Caligiore, D., Pezzulo, G., Baldassarre, G., Bostan, A. C., Strick, P. L., Doya, K., ... & Herreros, I. [2017b], *Consensus Paper: Towards a Systems-Level View of Cerebellar Function: The Interplay Between Cerebellum, Basal Ganglia, and Cortex*, The Cerebellum, 16, n. 1, pp. 203–229.

Caligiore, D., Silvetti, M., D'Amelio, M., Puglisi-Allegra, S. & Baldassarre, G. [2020], *Computational Modeling of Catecholamines Dysfunction in Alzheimer's Disease at Pre-Plaque Stage*, Journal of Alzheimer's Disease, 77, n. 1, pp. 275–290.

Cerasa, A., Gaggioli, A., Marino, F., Riva, G. & Pioggia, G. [2022], *The Promise of The Metaverse In Mental Health: The New Era of MEDverse*, Heliyon, 8, n. 11, p. e11762.

Clark, A [2014], Perceiving as Predicting, in D. Stokes, M. Matthen & S. Biggs (eds), *Perception and Its Modalities*. Oxford University Press, New York, pp. 23–43.

De Vugt, M. E. & Verhey, F. R. [2013], *The Impact of Early Dementia Diagnosis and Intervention on Informal Caregivers*, Progress in Neurobiology, 110, pp. 54–62.

Dirkx, M. F., den Ouden, H., Aarts, E., Timmer, M., Bloem, B. R., Toni, I. & Helmich, R. C. [2016], *The Cerebral Network of Parkinson's Tremor: An Effective Connectivity fMRI Study*, Journal of Neuroscience, 36, n. 19, pp. 5362–5372.

Dukart, J., Sambataro, F. & Bertolino, A. [2016], *Accurate Prediction of Conversion to Alzheimer's Disease Using Imaging, Genetic*, and Neuropsychological Biomarkers, Journal of Alzheimer's Disease, 49, n. 4, pp. 1143–1159.

Dy, M., Olazo, K., Lisker, S., Brown, E., Saha, A., Weinberg, J. & Sarkar, U. [2023], *Virtual Reality for Chronic Pain Management among Historically Marginalized Populations: Systematic Review of Usability Studies*, Journal of Medical Internet Research, 25, p. e40044.

Etgen, T. [2015], *Kidney Disease as a Determinant of Cognitive Decline and Dementia*, Alzheimer's Research & Therapy, 7, n. 1, pp. 1–7.

Eyigoz, E., Mathur, S., Santamaria, M., Cecchi, G. & Naylor, M. [2020], *Linguistic Markers Predict Onset of Alzheimer's Disease*, EClinicalMedicine, 28, pp. 1–9.

Farinella, F., Merone, M., Bacco, L., Capirchio, A., Ciccozzi, M. & Caligiore, D. [2022], *Machine Learning Analysis of High-Grade Serous Ovarian Cancer Proteomic Dataset Reveals Novel Candidate Biomarkers*, Scientific Reports, 12, n. 1, p. 3041.

Frias, C. E., Cabrera, E. & Zabalegui, A. [2020], *Informal Caregivers' Roles in Dementia: The Impact on Their Quality of Life*, Life, 10, n. 11, p. 251.

Grassi, M., Rouleaux, N., Caldirola, D., Loewenstein, D., Schruers, K., Perna, G. & Michel Dumontier, M. [2019], *A Novel Ensemble-Based Machine Learning Algorithm to Predict the Conversion from Mild Cognitive Impairment to Alzheimer's Disease Using Socio-Demographic Characteristics, Clinical Information, and Neuropsychological Measures*. Frontiers in Neurology, 10, 756, doi: https://doi.org/10.3389/fneur.2019.00756.

Grueso, S. & Viejo-Sobera, R. [2021], *Machine Learning Methods for Predicting Progression from Mild Cognitive Impairment to Alzheimer's Disease Dementia: A Systematic Review*, Alzheimer's Research & Therapy, 13, pp. 1–29.

Hardy, J. [2006], *Alzheimer's Disease: The Amyloid Cascade Hypothesis: An Update And Reappraisal*, Journal of Alzheimer's Disease, 9, n. 3, pp. 151–153.

Helmich, R. C. [2018], *The Cerebral Basis of Parkinsonian Tremor: a Network Perspective*, Movement Disorders, 33, n. 2, pp. 219–231.

Huynh-The, T., Pham, Q. V., Pham, X. Q., Nguyen, T. T., Han, Z. & Kim, D. S. [2023], *Artificial Intelligence for the Metaverse: A Survey*, Engineering Applications of Artificial Intelligence, 117, p. 105581.

Huys, Q. J., Maia, T. V. & Frank, M. J. [2016], *Computational Psychiatry as a Bridge from Neuroscience to Clinical Applications*, Nature Neuroscience, 19, n. 3, pp. 404–413.

Jacobs, H. I., Hopkins, D. A., Mayrhofer, H. C., Bruner, E., van Leeuwen, F. W., Raaijmakers, W. & Schmahmann, J. D. [2018], *The Cerebellum in Alzheimer's Disease: Evaluating Its Role in Cognitive Decline*, Brain, 141, n. 1, pp. 37–47.

Jakhar, D. & Kaur, I. [2020], *Current Applications of Artificial Intelligence for COVID-19*, Dermatologic Therapy, doi: https://doi.org/10.1111/dth.13654.

Jones, O. T., Matin, R. N., van der Schaar, M., Bhayankaram, K. P., Ranmuthu, C. K. I., Islam, M. S., … & Walter, F. M. [2022], *Artificial Intelligence and Machine Learning Algorithms for Early Detection of Skin Cancer in Community and Primary Care Settings: A Systematic Review*, The Lancet Digital Health, 4, n. 6, pp. e466–e476.

Juhola, M., Joutsijoki, H., Penttinen, K., & Aalto-Setälä, K. (2018). *Detection of Genetic Cardiac Diseases by Ca2+ Transient Profiles Using Machine Learning Methods*. Scientific Reports, 8 (1), 9355. Article 9355.

Keshavarzi Arshadi, A., Webb, J., Salem, M., Cruz, E., Calad-Thomson, S., Ghadirian, N., … & Yuan, J. S. [2020], *Artificial Intelligence for COVID-19 Drug Discovery and Vaccine Development*, Frontiers in Artificial Intelligence, 65, pp. 1–13.

Liang, H., Tsui, B. Y., Ni, H., Valentim, C. C., Baxter, S. L., Liu, G. & Xia, H. [2019], *Evaluation and Accurate Diagnoses of Pediatric Diseases Using Artificial Intelligence*, Nature Medicine, 25, n. 3, pp. 433–438.

Liu, M., Cheng, D., Wang, K. & Wang, Y. [2018], *Multi-Modality Cascaded Convolutional Neural Networks for Alzheimer's Disease Diagnosis*, Neuroinformatics, 16, pp. 295–308.

Long, X., Chen, L., Jiang, C. & Zhang, L. [2017], *Prediction and Classification of Alzheimer Disease Based on Quantification of MRI Deformation*, PLOS One, 12, n. 3, p. e0173372.

Luukinen, H., Viramo, P., Koski, K., Laippala, P. & Kivelä, S. L. [1999], *Head Injuries and Cognitive Decline among Older Adults: A Population-Based Study*, Neurology, 52, n. 3, pp. 557–557.

Merone, M., D'Addario, S. L., Mirino, P., Bertino, F., Guariglia, C., Ventura, R., ... & Caligiore, D. [2022], *A Multi-Expert Ensemble System for Predicting Alzheimer Transition Using Clinical Features*, Brain Informatics, 9, n. 1, pp. 1–11.

Mesulam, M. M. [2004], *The Cholinergic Lesion of Alzheimer's Disease: Pivotal Factor or Side Show?* Learning & Memory, 11, n. 1, pp. 43–49.

Moustafa, A. A. [2021], *Alzheimer's Disease: Understanding Biomarkers, Big Data Alzheimer's Disease: Understanding Biomarkers, Big Data, and Therapy, Londra*, Academic Press.

Myszczynska, M. A., Ojamies, P. N., Lacoste, A. M., Neil, D., Saffari, A., Mead, R., ... & Ferraiuolo, L. [2020], *Applications of Machine Learning to Diagnosis and Treatment of Neurodegenerative Diseases*, Nature Reviews Neurology, 16, n. 8, pp. 440–456.

Nobili, A., Latagliata, E. C., Viscomi, M. T., Cavallucci, V., Cutuli, D., Giacovazzo, G., ... & D'Amelio, M. [2017], *Dopamine Neuronal Loss Contributes to Memory and Reward Dysfunction in a Model of Alzheimer's Disease*, Nature Communications, 8, n. 1, pp. 1–14.

Norton, S., Matthews, F. E., Barnes, D. E., Yaffe, K. & Brayne, C. [2014], *Potential for Primary Prevention of Alzheimer's Disease: An Analysis of Population-Based Data*, The Lancet Neurology, 13, n. 8, pp. 788–794.

Parisi, D. [2001], *Simulazioni: la Realtà Rifatta nel Computer*, Bologna, Il Mulino.

Park, S. & Kim, S. [2022], *Identifying World Types to Deliver Gameful Experiences For Sustainable Learning in the Metaverse*, Sustainability, 14, n. 3, p. 1361.

Paulin, M. G. [1993], *The Role of the Cerebellum in Motor Control and Perception*, Brain, Behavior and Evolution, 41, n. 1, pp. 39–50.

Pistollato, F., Sumalla Cano, S., Elio, I., Masias Vergara, M. & Giampieri, F. [2016], *Role of Gut Microbiota and Nutrients in Amyloid Formation and Pathogenesis of Alzheimer Disease*, Nutrition Reviews, 74, n. 10, pp. 624–634.

Plassman, B. L., Havlik, R. J., Steffens, D. C., Helms, M. J., Newman, T. N., Drosdick, D., ... & Breitner, J. C. S. [2000], *Documented Head Injury in Early Adulthood and Risk of Alzheimer's Disease and Other Dementias*, Neurology, 55, n. 8, pp. 1158–1166.

Platero, C., Lin, L. & Tobar, M. C. [2019], *Longitudinal Neuroimaging Hippocampal Markers for Diagnosing Alzheimer's Disease*, Neuroinformatics, 17, n. 1, pp. 43–61.

Pruntel, S. M., van Munster, B. C., de Vries, J. J., Vissink, A. & Visser, A. [2024], *Oral Health as a Risk Factor for Alzheimer Disease*, The Journal of Prevention of Alzheimer's Disease, 11, n. 1, pp. 249–258.

Rasmussen, J. & Langerman, H. [2019], *Alzheimer's Disease-Why We Need Early Diagnosis*, Degenerative Neurological and Neuromuscular Disease, 9, pp. 123–130.

Rawat, V., Joshi, S., Gupta, S., Singh, D. P. & Singh, N. [2022], *Machine Learning Algorithms for Early Diagnosis of Diabetes Mellitus: A Comparative Study*, Materials Today: Proceedings, 56, pp. 502–506.

Riva, G., Di Lernia, D., Sajno, E., Sansoni, M., Bartolotta, S., Serino, S., ... & Wiederhold, B. K. [2021], *Virtual Reality Therapy in the Metaverse: Merging VR for the Outside With VR for the Inside*, Annual Review of Cybertherapy & Telemedicine, 19, pp. 3–8.

Riva, G. & Wiederhold, B. K. [2022], *What the Metaverse Is (Really) and Why We Need to Know About It*, Cyberpsychology, Behavior, and Social Networking, 25, n. 6, pp. 355–359.

Schmahmann, J. D. & Caplan, D. [2006], *Cognition, Emotion and the Cerebellum*, Brain, 129, n. 2, pp. 290–292.

Schwartenbeck, P. & Friston, K. [2016], *Computational Phenotyping in Psychiatry: A Worked Example*. eNeuro, 3, n. 4, doi: https://doi.org/10.1523/ENEURO. 0049-16.2016.

Spiegel, B., Fuller, G., Lopez, M., Dupuy, T., Noah, B., Howard, A., ... & Danovitch, I. [2019], *Virtual Reality for Management of Pain in Hospitalized Patients: A Randomized Comparative Effectiveness Trial*, PLOS One, 14, n. 8, p. e0219115.

Srinivasu, P. N., Shafi, J., Krishna, T. B., Sujatha, C. N., Praveen, S. P. & Ijaz, M. F. [2022], *Using Recurrent Neural Networks for Predicting Type-2 Diabetes from Genomic and Tabular Data*, Diagnostics, 12, n. 12, pp. 1–30.

Testi, S., Peluso, S., Fabrizi, G. M., Antenora, A., Russo, C. V., Pappatà, S., ... & Filla, A. [2014], *A Novel PSEN1 Mutation in a Patient with Sporadic Early-Onset Alzheimer's Disease and Prominent Cerebellar Ataxia*, Journal of Alzheimer's Disease, 41, n. 3, pp. 709–714.

Vamathevan, J., Clark, D., Czodrowski, P., Dunham, I., Ferran, E., Lee, G., ... & Zhao, S. [2019], *Applications of Machine Learning in Drug Discovery and Development*, Nature Reviews Drug Discovery, 18, n. 6, pp. 463–477.

Wani, S. U. D., Khan, A., Thakur, G., Gautam, S. P., Ali, M., Alam, P., ... & Shakeel, F. [2022], *Utilization of Artificial Intelligence in Disease Prevention: Diagnosis, Treatment, and Implications for the Healthcare Workforce*, Healthcare, 10, n. 4, pp. 1–17.

Whiteneck, G. G., Gerhart, K. A. & Cusick, C. P. [2004], *Identifying Environmental Factors that Influence the Outcomes of People with Traumatic Brain Injury*, The Journal of Head Trauma Rehabilitation, 19, n. 3, pp. 191–204.

Wilson, H., Dervenoulas, G., Pagano, G., Koros, C., Yousaf, T., Picillo, M., ... & Politis, M. [2019], *Serotonergic Pathology and Disease Burden in the Premotor and Motor Phase of A53T α-Synuclein Parkinsonism: A Cross-Sectional Study*, The Lancet Neurology, 18, n. 8, pp. 748–759.

Wu, T. & Hallett, M. [2013], *The Cerebellum in Parkinson's Disease*, Brain, 136, n. 3, pp. 696–709.

CHAPTER 3

Ethics, Privacy, and Other Issues in AI Medicine

3.1 WILL AI REPLACE DOCTORS?

The question of whether the advancement of artificial intelligence (AI)-related technologies and the spread of automation threaten employment, particularly in the medical field, has been raised for years [Acemoglu and Restrepo 2020]. Historically, automation technologies focused mainly on replacing repetitive, labor-intensive, or dangerous jobs, such as those on assembly lines or in the petrochemical industry. Today, however, the landscape is changing. In addition to these established applications of AI, new ones are emerging that involve automating jobs that are not strictly repetitive but require high levels of specialization and creativity, such as teaching, law, journalism, and medicine. This shift raises concerns that advances in AI might reduce job opportunities. But is this the case? Some research suggests otherwise, proposing that the development and spread of AI will *transform* the labor market, with workers needing new skills [Bessen 2016; Frank et al. 2019]. In my book *IA istruzioni per l'uso*, I discuss how research from the Artificial Intelligence Observatory at the School of Management of the Politecnico di Milano, the American company Cognizant, and the World Economic Forum shows that this transformation could lead to the emergence of new jobs in the future [Caligiore 2022].

AI will undoubtedly transform the labor market, including the medical sector. A recent study by Goldman Sachs, one of the largest investment banks[1] in the world, estimates that between 26% and 28% of jobs in the US healthcare sector could become automated, depending on the role. This shift largely stems from the rise of generative AI – AI that can autonomously create new content, such as images, text, sound, and video, in ways similar to human creativity. Such software can engage in conversations, supplying complex and detailed responses on any topic, and can significantly support doctors and other healthcare professionals. However, the impressive capabilities of these technologies raise concerns, one of the most pressing being: Will they replace healthcare personnel?

ChatGPT serves as a prime example of generative AI. It is a chatbot – software designed to engage in conversations with users. While chatbots have existed for at least a decade, ChatGPT offers something revolutionary. OpenAI developed ChatGPT, which uses a powerful AI model known as a *transformer* (ChatGPT stands for Chat Generative Pre-trained Transformer). This artificial neural network processes long sequences of words simultaneously. Traditional artificial neural networks process data sequentially, with each step depending only on the immediately preceding data. In contrast, the transformer uses an *attention mechanism* to capture long-term relationships between words more efficiently and accurately. Traditional models struggle to identify connections between distant words. By contrast, transformer models analyze the entire sequence, enabling them to highlight relevant semantic links between words, even when they are far apart. Transformers have proven effective in translation, text generation, and speech synthesis. ChatGPT quickly gained great popularity, attracting one million users within a week of its release on 22 November 2022. Within just a few months, its user base surpassed 100 million worldwide. This chatbot has become the fastest-growing application in the tech industry, setting a new benchmark for user numbers. Box 3.1 shows an example of how users can interact with ChatGPT.

BOX 3.1   THE INTERACTION BETWEEN A HUMAN USER AND CHATGPT, SPECIFICALLY GPT-3.5. GPT-3.5 IS A CHATGPT FREE VERSION DEVELOPED BY OPENAI, TRAINED WITH DATA COLLECTED UP TO JANUARY 2022. THE CHATBOT AUTONOMOUSLY GENERATES MULTIPLE RESPONSES TO THE USER QUERY.

Human user question. I am 55 years old and have type 2 diabetes. In the past month, I have noticed an increase in my blood pressure. I am currently taking metformin for diabetes. What treatment options for blood pressure might be appropriate for me?

GPT-3.5 answer. It's important to manage both blood sugar levels and blood pressure, especially since having diabetes increases the risk of developing hypertension (high blood pressure). If you've noticed a rise in your blood pressure, there are several approaches you could consider. However, you should always consult with your healthcare provider before making any changes to your treatment plan. Here are some general options that might be appropriate:

1. Lifestyle Changes:
   - Diet: A heart-healthy diet, like the DASH (Dietary Approaches to Stop Hypertension) diet, which focuses on reducing sodium intake, increasing potassium, and eating more fruits, vegetables, whole grains, and lean proteins, can help lower blood pressure.
   - Exercise: Regular physical activity, such as 30 minutes of moderate exercise (walking, swimming, cycling) most days of the week, can help reduce both blood sugar and blood pressure.
   - Weight management: If you are overweight or obese, losing even a small amount of weight can help lower blood pressure and improve overall health.
   - Stress management: Techniques like meditation, yoga, or deep breathing can help reduce stress and potentially lower your blood pressure.
2. Medications for High Blood Pressure:
   Since you already take metformin for diabetes, your doctor might consider medications that are both safe for people with diabetes and effective for lowering blood pressure. Some commonly used classes include:
   - ACE inhibitors (e.g., enalapril, lisinopril): These are often preferred for people with diabetes because they help protect kidney function, which can be a concern for diabetics. ACE inhibitors can also lower blood pressure.
   - Angiotensin II receptor blockers (ARBs) (e.g., losartan, valsartan): Similar to ACE inhibitors, ARBs can help lower blood pressure and are also kidney-friendly for people with diabetes.
   - Calcium channel blockers (e.g., amlodipine): These can help lower blood pressure by relaxing blood vessels, and they're commonly used in combination with other medications.
   - Thiazide diuretics (e.g., hydrochlorothiazide): These can help reduce blood pressure by removing excess salt and fluid through urination. However, they should be used cautiously with people with diabetes, as they can increase blood sugar in some cases.
   - Beta-blockers (e.g., atenolol): These might be used if there's a concern about heart health as well as high blood pressure. However, they can sometimes interfere with blood sugar management, so their use must be carefully monitored.

3. Monitor Blood Pressure Regularly:
Keeping track of your blood pressure at home can help you and your healthcare provider understand how well your treatment plan is working. Many pharmacies also offer blood pressure machines for use in-store or online.

4. Consider the Impact of Metformin:
Metformin primarily affects blood sugar, but it can also slightly influence blood pressure. Make sure to discuss whether any changes in your diabetes medication might be necessary to manage both conditions effectively.

5. Regular Check-ups:
It's crucial to have regular check-ups with your healthcare provider to ensure that both your blood pressure and blood sugar are well-managed. Monitoring kidney function and other related health markers is especially important when managing both conditions.

Ultimately, managing high blood pressure and diabetes together requires a comprehensive approach. Your doctor may choose one or a combination of treatments tailored to your specific needs and health status. Please reach out to them to discuss your symptoms and work on a personalized plan.

ChatGPT is a large language model (LLM) because it has been trained on vast amounts of text data to learn how a given language uses words and phrases. This feature enables it to answer questions, perform language tasks, and engage in conversations. Specifically, ChatGPT is a generalist LLM, meaning it can generate text on a wide range of topics, including the medical field. In contrast, some chatbots focused on specific sectors, such as BioGPT, the medical version of ChatGPT, or MedPalm, a transformer with 540 billion parameters. MedPalm, developed by researchers at Google and DeepMind, can answer a broad range of medical questions. In some cases, these generative AI systems have shown a high success rate in providing correct answers to medical questions, with performance levels comparable to those of human experts. However, this does not mean that an LLM can perform hypothetical-deductive reasoning, a crucial skill in the medical field for making differential diagnoses.[2] Instead, LLMs excel at identifying correlations and have an impressive memory capacity, which allows them to process and generate large amounts of information [Liévin, Hother and Winther 2022]. In the future, these types of software could have various applications. For example, they could be used for medical training, working as a virtual patient to answer questions about health conditions. They could also assist doctors by suggesting potential therapeutic options based on the specific characteristics of a patient. To date, there are no AI generative-based applications certified as medical devices. Still, the likelihood of such applications emerging in the near future is high. For example, a demo version of an app called Glass AI[3] recently launched, which uses an AI model to formulate a clinical plan[4] based on a clinical case description.

In general, generative AI, such as ChatGPT or MedPalm, can allow us to obtain suggestions, insights, and accurate answers on several topics. However, it is crucial to approach generative AI with a critical mindset. Box 3.2 outlines some recommendations for the safe and responsible use of generative AI.

BOX 3.2   TIPS FOR USING GENERATIVE AI EFFECTIVELY AND RESPONSIBLY

Verify the accuracy of responses. It is essential to verify the accuracy of the information provided by generative AI by cross-referencing it with reliable and authoritative sources. Sometimes, generative AI can produce "hallucinations" – inaccurate or false information, often citing nonexistent sources. These hallucinations can occur for several reasons. For instance, AI may provide incorrect responses if the user has not provided a clear context. Additionally, if the data used to train the AI contains ambiguous, contradictory, or distorted information, the AI may generate responses reflecting these inconsistencies. Moreover, if the AI has been trained on a narrow dataset without exposure to a wide variety of cases, it may overfit – becoming too reliant on the training data and struggling to adapt to new or different information. In general, AI hallucinations remind us that we are dealing with a tool that lacks the awareness and understanding that humans possess.

Ask clear and varied questions. Asking clear and specific questions with well-defined context is crucial to minimize the risk of receiving imprecise answers from generative AI. Additionally, it can be helpful to regenerate multiple responses to the same question (as generative AI often produces different answers even when the question remains the same) or to rephrase the question in various ways. This approach allows you to explore the topic from different perspectives and gain a deeper understanding.

Use for repetitive tasks. Generative AI can be a valuable tool for handling repetitive tasks, thus saving time. For example, ChatGPT can generate drafts, complete standard forms, or perform tasks that require minimal creativity. This type of use can assist professionals, such as doctors, with administrative and bureaucratic work, where producing text with a standardized structure is often necessary.

Continue using traditional approaches. It is essential to continue utilizing traditional sources to study a problem, such as books, scientific articles, and attending conferences. Staying tuned with established methods is crucial to avoid becoming overly reliant on a single source – generative AI – and to discover new and original ways of solving problems. These approaches are often difficult to uncover if one relies solely on AI-generated information, which typically reflects the majority viewpoint. Sometimes, considering *individual perspectives*, even those that differ from the mainstream *can make a significant difference* in problem-solving.

Generative AI to stimulate creativity. Consider generative AI answers as a starting point, not the final word. They can serve as a stimulus to explore other creative avenues and alternative solutions beyond those proposed by AI. For instance, if we ask ChatGPT about the typical symptoms of Parkinson's disease, we might receive a list of common symptoms such as rest tremor, muscle rigidity, bradykinesia (slowness of movement), postural instability, difficulty walking, balance issues, decreased sense of smell, sleep disturbances, and depression. While these are the typical symptoms, we can use this information as a springboard to explore rare or atypical symptoms of Parkinson's disease that AI might not mention, as it tends to reflect the majority viewpoint. By considering recent research and individual perspectives (as noted in the previous point), we could delve into the connections between Parkinson's disease and symptoms of other neurodegenerative conditions, like Alzheimer's disease It is [Caligiore, Giocondo and Silvetti 2022]. Additionally, we could explore how Parkinson's disease affects an individual's daily life, including the ability to engage in creative activities [Lauring et al. 2019]. Another example: If we ask ChatGPT about common therapies for Parkinson's disease, we might receive a summary of the typical treatments, such as medication for symptom management (e.g., tremors and muscle stiffness), physical therapy to improve mobility and strength, and deep brain stimulation for cases where medication is ineffective. Again, AI responses serve as a starting point. Building on recent research, we could also investigate the potential benefits of complementary and alternative therapies, like art therapy and music therapy, in managing Parkinson's symptoms [Ettinger et al. 2023; Machado Sotomayor et al. 2021]. Exploring these alternatives could lead to fresh perspectives and innovative ideas for addressing Parkinson's disease.

Using generative AI at the right time. Finally, it is important to remember that to use generative AI effectively and responsibly, one should first develop essential skills, such as writing and critical thinking. These skills are best acquired through traditional education, without relying on such technologies. Generative AI can be most beneficial at a later stage in a person's education and intellectual development. In general, introducing generative AI too early – such as in primary and early secondary school – could pose risks to cognitive and cultural growth.

AI should be viewed as a tool, not a colleague, and it should always be seen as a support, never a replacement. It is important to find the right balance in using AI tools, both for doctors and patients. For instance, patients can use AI to learn more about their health, such as what diseases may be associated with their symptoms or what lifestyle changes might help prevent illness. However, this does not mean AI should replace doctors. Instead, it means patients can come to their healthcare provider with a better understanding of their health, allowing for more informed discussions. Generative AI synthesizes information from the web to answer questions posed by users, whether doctors or patients. Regarding scientific article writing, some prestigious journals have banned the use of generative AI in the editorial process, even employing tools to detect if a text was written by an AI model. Conversely, other journals have established guidelines for authors, reviewers, and editors to ensure the responsible use of AI tools. Once again, AI should not be seen as a colleague that replaces us in writing scientific articles but as a tool that helps us communicate our research findings more clearly and effectively.

Generative AI can assist researchers in quickly summarizing key findings from previous studies and the current state of the art, allowing them to consider all relevant aspects that colleagues have explored in prior research. This feature helps to build upon the body of scientific work that inspired and motivated new investigations. Aside from researchers, clinicians can also benefit from generative AI by rapidly obtaining critical information from scientific articles – information that would otherwise require significant time to extract manually by reading and summarizing multiple papers. This capability can improve efficiency, accuracy, and speed of clinical decisions, allowing doctors to deliver better patient care. For example, a physician could ask generative AI a specific question about the results of clinical trials for a particular drug or treatment and quickly receive a summary of the key findings. Alternatively, they could ask AI to locate the most recent studies or guidelines on a specific medical treatment. Generative AI can swiftly scan extensive scientific literature and provide the most relevant and up-to-date information to help physicians make informed decisions.

In some cases, introducing partial physician substitution through AI could offer significant benefits. For example, the administrative burdens on doctors – such as patient registration, updating medical records, and managing billing – can be heavy. AI can automate many of these tasks, reducing the physician workload. AI systems can handle patient registration, ensuring accurate data entry. They can also automatically update medical records. AI can streamline the billing process, minimizing errors and speeding up processing times. This automation not only frees up valuable physician time, allowing them to focus more on patient care, but also improves the overall efficiency of the healthcare system. By reducing administrative tasks, doctors can spend more time on human interactions and clinical decisions, enhancing the quality of healthcare. Patients also benefit from quicker, more accurate information management, which could help reduce waiting times. Ultimately, using AI to automate administrative procedures in healthcare leads to more efficient and accurate information management, benefiting both physicians and patients.

AI can provide answers primarily by evaluating medical data. In contrast, human healthcare workers can assess not only medical data but also personal factors that may affect a patient's health. Additionally, healthcare professionals have the unique ability to offer empathy, understanding, and emotional support, all of which are crucial for building a trusting doctor-patient relationship. While healthcare professionals must develop technical skills to use AI tools, they should also nurture and enhance qualities that AI cannot replicate – such as soft skills and emotional intelligence. These human traits are essential for introducing an emotional component into work environments (not just in healthcare), where AI can handle the more mechanical tasks, and humans can focus on the more relational and affective aspects. To prepare for the AI revolution, training programs for healthcare professionals should include courses in psychology and soft skills (see Chapter 4, Section 4.4). Undoubtedly, the role of physicians will evolve as AI becomes more integrated into healthcare, but *the need for human interactions* will remain central to patient care.

3.2 WILL WE LOSE CONTROL OVER OURSELVES?

In the 1970s, the Italian television network RAI aired the famous TV drama *Gamma*, which explored brain transplantation and its profound ethical implications. The drama presented brain transplantation as if it were on par with heart or other vital organ transplants. While we now regard this idea as pure fantasy, the thought-provoking questions raised by the script about how brain transplantation could affect personality offer a fascinating and plausible perspective. Replacing the brain would essentially mean replacing an individual's personality, making it an extremely invasive method of manipulating one's thought processes. Today, rather than transplanting the entire brain, scientific research focuses more on replacing specific malfunctioning parts of the brain or modulating brain functions through neurostimulation [Mattioli et al. 2023]. These approaches present a more realistic – and perhaps, within certain ethical limits we will discuss later – a more ethically sustainable perspective.

In 2019, a team of researchers at Columbia University used laser technology to control the activity of a group of neurons in the visual cortex of mice, manipulating the animals perceptual abilities from the outside. Specifically, after the laser manipulation, the mice saw a symbol that was not present in reality but which they had previously associated with the presence of water. As a result, the mice began licking at a water container as if it contained water, even though neither the symbol nor water was present in their environment [Carrillo-Reid et al., 2019]. The research coordinator, neuroscientist Rafael Yuste, emphasized that the neurotechnology used in their experiment could have revolutionary applications in the years ahead. On one hand, this technology could radically transform the treatment of neurodegenerative diseases, such as Alzheimer's or Parkinson's disease, by enabling noninvasive interventions to restore typical neural activity in compromised brain areas. On the other hand, Yuste also raised a concern with critical ethical implications: The potential to use such technologies to artificially select our sensations as if from a "sensory menu". For example, we could experience the pleasure of tasting our favorite food without actually eating it. These reflections led Rafael Yuste and other prominent scientists to establish the *Neurorights Initiative*,[5] an international group of experts in neuroscience and law that developed a list of fundamental rights in the field of neuroscience, known as "neurorights" [Yuste et al. 2017]. Among these is the *right to mental privacy*, which aims to protect brain data related to our thoughts. Currently, functional magnetic resonance imaging (fMRI) enables the real-time monitoring of neuronal activity in different brain regions as a person performs specific tasks. Combined with AI techniques, this technology could soon support the direct reading of human thoughts. The team of Japanese researchers led by Professor Yukiyasu Kamitani made a pioneering step in this direction. Their work demonstrated the possibility of using deep neural networks to reconstruct the image a person is looking at simply by examining their brain activity [Shen et al. 2017].

Another neuro-right that needs protection is the *right to the integrity of personal identity*, which aims to prevent the connection of electronic devices (e.g., computers) to the brain from affecting an individual's sense of identity or free will. The Neuralink project,[6] conceived and supported by entrepreneur Elon Musk, seeks to develop new neurotechnologies that enhance the capabilities of the human mind through AI. Using sophisticated machine learning algorithms, the system will identify which brain neurons to stimulate to improve cognitive abilities. According to Musk, this technology will not only allow more precise control of external devices, such as exoskeletons and computers but also enable information to be transferred directly into our brains. This could lead to accelerated learning, allowing us to acquire knowledge and skills with minimal effort. However, it also raises the possibility that, in some situations, an external device like a computer could directly influence how the brain makes decisions. It will become nearly impossible to distinguish whether brain activity results from external manipulation (such as a computer or AI) or the brain's autonomous physiological processes.

Neurostimulation techniques, which have been used for several years in the medical and scientific fields to modulate brain activity – such as transcranial magnetic stimulation (TMS) and transcranial direct current stimulation (tDCS)[7] – raise an important question: Is it clear whether external brain stimulation can only influence thoughts and perceptions that we are consciously aware of, or if it can also affect thoughts and mental processes that occur outside our conscious awareness (unconscious contents)? This question raises ethical concerns, as it could challenge the concept of free will, making it difficult to determine whether a thought results from external influences (such as TMS or tDCS stimulation) or the outcome of autonomous personal decisions. The future development of increasingly sophisticated neurotechnologies, such as those proposed by Neuralink, could significantly impact our understanding of how the mind works and individual autonomy. Our thoughts and behavior result from a complex interaction between rational, conscious drives and irrational, unconscious forces. This also includes automatic mental processes, such as dreaming, as well as involuntary lapses or errors [Freud 1999]. The unconscious mind regulates processes such as perception, learning, memory, and emotions, managing procedural memory that controls automatic actions like walking or cycling. This aspect, known as the 'cognitive unconscious', enables us to perform tasks without being aware of the details of our movements, such as catching a ball mid-air. However, there is also a deeper layer of the unconscious responsible for managing experiences that are too intense or dangerous. The mind creates defense mechanisms that isolate such experiences in unconscious memory, forming the "removed unconscious". Our thinking, decisions, and behavior depend on a delicate balance between conscious awareness and unconscious processes. Throughout evolution, the brain learned to maintain this balance in environments without external neurostimulation. External stimulation could disrupt this balance, allowing unconscious mechanisms – operating beyond our voluntary control – to override rational thought. This process could eventually lead to a new form of consciousness controlled externally rather than by our own will. The consequences of these dynamics on our minds remain unpredictable [Legrenzi and Umiltà 2018].

Cognitive enhancement through neural stimulation could alter our perception of personal limits, potentially leading us to lose control over ourselves. A study conducted by researchers in the American Armed Forces demonstrated that electrical stimulation of specific brain areas can improve our ability to multitask effectively [Nelson et al. 2016]. The advancement of machine learning techniques in interpreting cerebral electrical signals has further increased the effectiveness of these neurotechnologies. However, this approach prioritizes speed – the ability to quickly decide how to handle various situations – over the slower processes often necessary for developing argumentation, critical reasoning, and thoughtful decision-making. It creates the illusion that we have no limits in our decision-making. Yet, the concept of limits is essential in our lives; it is a crucial tool for avoiding the flattening of knowledge, which can result from the overuse of hybrid technologies [Caligiore 2022]. Recognizing and respecting our limits plays a critical role in maintaining self-control. When we acknowledge our limitations, we are more likely to make informed decisions, set realistic goals, evaluate the consequences of our actions, and adopt responsible behaviors. Believing we have no limits can lead to an egocentric, narcissistic mindset, undermining interpersonal relationships and emotional well-being while making it harder to consider the needs and rights of others. Although *learning from our limitations* may take longer, it *values not just the result but also the process* followed to achieve it. Instead of seeing limits as obstacles to overcome, we should view them as boundaries that define a space for knowledge. The differences between us and others, defined by individual boundaries, are a strength in our interactions, fostering the development of knowledge, critical thinking, and meaningful dialogue. In addition to influencing our sense of limits, the use of hybrid technologies that combine AI and neurotechnologies to manipulate brain activity may give rise to new forms of discrimination, particularly against those who choose not to enhance their cognitive abilities with these tools and prefer to explore the world through their limitations. Therefore, it is essential to critically examine the social and ethical implications of these technologies, ensuring equal access to knowledge while promoting a genuine and profound understanding in our learning processes, problem-solving, and decision-making.

In addition to emerging neurotechnologies, the metaverse could also influence our ability to maintain self-control. Immersing ourselves in a space where the boundaries between the real and virtual worlds blur may subtly encourage impulsive and obsessive behavior. In such a fluid environment, people may be more prone to irrational and compulsive actions fueled by a sense anonymity and the absence of immediate real-world consequences. For example, in an online gaming environment, an individual might be tempted to verbally attack other players without considering the emotional or social impact of their actions in real life. Similarly, in the metaverse, someone might feel emboldened to offer medical advice without proper qualifications, potentially endangering others health. This impulsive behavior often arises from a sense of emotional detachment and the perceived safety of the virtual world, but it can lead to serious real-world consequences. The phenomenon is also present in social media addiction, where the constant pursuit of approval and instant gratification drives obsessive behaviors, such as repeatedly checking for likes and comments.

The concept of a perfect world, as often envisioned in the metaverse, raises several concerns and risks in the medical field. If people spend most of their time in the metaverse, reduced physical activity could lead to muscle atrophy, poor blood circulation, and other health issues associated with a sedentary lifestyle. Additionally, a metaverse "perfect world" could contribute to social isolation and depression, as limiting real-world social interactions may worsen mental health conditions like anxiety. People might lose essential social skills, such as managing conflict, practicing compassion, and developing empathy – all crucial for healthy relationships and mental well-being. The absence of real-world challenges, defeats, and even disappointments could deprive individuals of valuable opportunities for personal growth. The metaverse could become a form of escape from reality, leading to addiction, where people neglect their real-life responsibilities, including self-care and health. Such an escape could result in issues like substance abuse, deterioration of physical and mental health, and other self-destructive behaviors. In the metaverse, people may create idealized digital versions of themselves, leading to a distortion of personal identity. The discrepancy between one virtual self and real self could cause psychological tension and self-esteem issues, ultimately undermining one's sense of control over their own identity.

3.3 IS RESPONSIBILITY WITH AI OR THE PHYSICIAN?

The widespread adoption of AI-related hybrid technologies in the medical field presents a fundamental dilemma: When a healthcare professional consults an AI for advice in making a decision, who is ultimately responsible for that decision? Should the responsibility fall on the healthcare provider, the AI developer, the vendor, or the AI itself? And if we were to assign responsibility to the AI, how would it be accountable for any costs or damages resulting from its decisions? Similar challenges arise with technologies like deep brain stimulation systems, such as the closed-loop devices used to treat neurological disorders like Parkinson's disease. These systems involve implanting electrodes in the brain, connected to a device known as a "cerebral pacemaker". This pacemaker delivers targeted electrical impulses to specific brain areas, helping regulate neuronal activity and alleviate symptoms. The term "closed-loop" refers to the system ability to autonomously adjust the intensity of the stimulation based on the patient's brain activity, optimizing the treatment in real-time. But if such a system causes harm, who is to be held accountable? The physician, the system designer, the patient, or the autonomous system itself? These scenarios raise profound questions about the nature of liability [Coeckelbergh 2016] and underscore the complex ethical and legal challenges posed by using AI and new technologies in healthcare. These challenges go beyond traditional concerns like privacy and data management (which remain relevant, as discussed in Section 3.4) and call for a clear legal framework. To ensure confidence in using AI in the medical field, it is crucial to establish a well-defined legal regime that assesses the impact of intelligent systems on clinical decision-making. The key challenge is understanding how to assign responsibility in cases of harm and developing a legal system capable of addressing these complex situations. Given recent advances in AI, particularly regarding the ethical considerations involved, it still seems premature to grant AI systems used in the medical field a form of "electronic personality". If AI were assigned such a status, it would be considered a legal entity capable of making autonomous decisions and thus held responsible for the consequences of those decisions, including any resulting harm. In this scenario, AI could be insured to cover the costs of any damages it causes. However, this concept raises complex legal and ethical questions that need further exploration.

Without legal personality for AI systems, assigning civil liability for damages caused by emerging AI-based technologies can follow two main approaches. First, responsibility may be assigned to those involved in the medical device creation and distribution under existing laws for machines and products. For AI, this involves identifying the liability of the manufacturer for marketing the intelligent system, the programmer who developed the AI algorithm, and the individual who trained the AI, often referred to as the trainer (who may also be the programmer). Second, the responsibility of end users – such as healthcare facilities, doctors, and patients – must also be considered. This creates a more complex scenario compared to traditional medical devices, as the legal liability framework can vary depending on the type of damage, the nature of the party responsible, and their relationship with the patient.

Here are some examples. In the case of a robot-assisted surgical procedure where a patient is injured, liability could fall on the manufacturer if a design or manufacturing defect caused the injury. The manufacturer is also liable if it fails to inform healthcare facilities and professionals about potential errors, contraindications, accident frequency, or the severity of possible harm. If the injury results from improper device use, liability may shift to the physician. For instance, if a healthcare provider uses robotic technology without proper training or chooses robotic surgery when traditional methods would be more appropriate for the patient condition, they could be held responsible. Healthcare facilities may also be held liable for not properly training staff, maintaining, or updating the AI system. In some cases, the patient could also be liable. Consider a heart patient using an AI-based device to monitor their health. If the patient misuses the device and ignores medical or manufacturer instructions, leading the AI to provide harmful advice, the patient might be responsible for any complications. Another scenario involves a doctor using an AI system to recommend medication doses or treatment duration. If the AI advice is incorrect, leading to harm, liability could fall on the manufacturer if it fails to warn about the system limitations with specific patient groups. Alternatively, the doctor may be at fault if they followed the AI recommendation without applying their clinical judgment. Healthcare professionals must ensure they are trained to use AI-based technologies and understand the system's capabilities and limitations. When using AI, doctors are responsible for verifying their recommendations based on their professional knowledge and the patient-specific condition, ensuring that the system suggestions are appropriate for the situation.

From a legal perspective, it is important to note that in Italy, healthcare professionals remain subject to the provisions of Law 24/2017

(Gelli-Bianco), which holds them liable for errors even when using new technologies or other medical equipment during healthcare procedures. The use of emerging technologies requires healthcare professionals to exercise caution, diligence, and competence. As mentioned earlier, if an accident results from a defect in technological equipment, the manufacturer is liable, along with others involved in the device production, in accordance with product liability regulations and the European Medical Device Regulation. The introduction of AI in healthcare could complicate identifying the cause of harm, making it difficult to determine whether the damage was due to a medical error or a technical malfunction. A clear example of this could arise from a complex diagnostic analysis performed by an AI-based system. Suppose a patient is misdiagnosed and harmed by incorrect treatment. In this case, it may be difficult for the patient to ascertain whether the error was due to the AI decision or a human mistake in interpreting the data provided by the system. Several measures should be implemented to address this issue. First, AI systems must be explainable, allowing healthcare professionals and patients to understand the decision-making process (see Section 3.5). Furthermore, whenever possible, all interactions with the system should be recorded and tracked for thorough analysis in case of incidents. Lastly, healthcare professionals should be trained not only in the technical use of AI systems but also in understanding their capabilities and limitations. Patients should also be educated in how these systems work. By implementing these measures, the problem of unknown causality in AI healthcare applications could be better managed, ensuring patient safety and accountability for all involved parties.

Insurance companies are exploring various solutions to address the complex liability issues surrounding AI in healthcare. One potential approach, already tested in Italy in similar contexts, involves adopting a *no-fault system*. This system would reduce legal disputes and streamline compensation without assigning blame to individuals or institutions. However, a key challenge remains: Who should cover the compensation costs without burdening public finances? A possible solution could involve a public-private partnership. For instance, insurance companies could collaborate with government institutions to establish a dedicated fund supported by voluntary contributions from healthcare companies and insurance providers. Additionally, tax incentives could encourage businesses to contribute to the fund, ensuring a financial response in case of a claim. An independent organization could manage the fund to ensure transparency, fairness, and efficiency in distributing compensation.

The collaboration between the public and private sectors could offer a sustainable financial model for managing liability in AI medical device use without burdening the public budget. However, from an ethical perspective, this approach may not fully address the issue. The lack of clear liability allocation – whether to the manufacturer, institution, physician, or patient – could discourage focus on essential aspects for the responsible use of AI in healthcare. For example, physicians might neglect proper training on using the device, manufacturers might fail to provide clear information on the benefits and risks of the system, or developers might rush AI training and reduce testing to save time. Essentially, this could create the impression that, even when mistakes occur, a common fund will ultimately cover the consequences, absolving those responsible of accountability. This is an argument that cannot be accepted.

3.4 DATA: PRIVACY, SECURITY, AND BIAS

The data are essential to train ML models. In the medical context, data often include sensitive patient information such as clinical test results, medical histories, genetic data, and diagnostic images. It is necessary to ensure ethical and secure use of this information, starting with patient privacy. In Europe, the General Data Protection Regulation (GDPR) Regulation (EU) 2016/679 provides comprehensive rules on personal data protection, including medical data. Under the GDPR, confidential data must be processed lawfully, fairly, transparently, and collected for specific, explicit, and legitimate purposes. Furthermore, data must be adequate, relevant, and used only for the specified purpose. When using patient data, it is crucial to anonymize it according to GDPR to protect the patient's identity and privacy. This aspect involves removing names, addresses, and any personally identifiable information. In some cases, explicit patient consent is necessary for processing sensitive data, such as health information. Other countries have similar laws (e.g., the Health Insurance Portability and Accountability Act (HIPAA) in the United States). Thus, to ensure the ethical and legal use of medical data for AI training, it is essential to comply with the GDPR or equivalent regulations in other countries and to obtain patients informed consent when necessary.

In addition to privacy in data management, another critical aspect is information security, especially data unauthorized access prevention. In digitized healthcare systems, patients EHRs (see Chapter 1, Section 1.2) must be encrypted using advanced algorithms. Encryption[8] ensures that only authorized personnel can access and update medical records, protecting the data from third parties. This action is essential for safeguarding sensitive information, such as medical histories, prescriptions, and test results. It also helps maintain patients trust in how the healthcare system manages their information.

Privacy and data security risks are also prevalent in virtual environments like the metaverse. This space offers numerous opportunities for virtual medical interactions, education, and procedure simulations, with patients using avatars to participate in virtual consultations or access online health services (see Chapter 2, Section 2.2). The data linked to these avatars – such as diagnostic, clinical, and financial information – are sensitive and must be carefully handled. *Virtual privacy* involves safeguarding these virtual identities and associated medical data to protect them from unauthorized access. In the medical metaverse, targeted cyberattacks and virtual identity theft of patients and doctors can occur, which may compromise the quality of care. Additionally, as discussed in Section 3.2, patients in virtual environments may be more likely to share sensitive information with healthcare providers or other users. These data could be misused if not adequately protected, endangering patient safety and privacy. Given the growing use of the medical metaverse, healthcare professionals and developers of virtual medical services must implement robust security measures. Advanced encryption protocols should be used to protect data, and users must be educated on the risks of oversharing personal information. Furthermore, clear and enforceable privacy and security regulations must be established to ensure the highest level of data protection. Only through these precautions can the medical metaverse become a trusted space for virtual care and interactions [Wang et al. 2022].

Finally, it is crucial to address the issue of bias [Pagano et al. 2023]. Consider an example: Suppose an algorithm is trained to diagnose a disease using historical data from a predominantly homogeneous group of patients, such as those of a specific ethnicity or gender. This could lead to the algorithm developing a bias, resulting in less accurate diagnoses for individuals outside of that group, which would severely undermine the quality of care. Therefore, identifying and mitigating bias in the data is essential before using it to train machine learning models, ensuring that the resulting algorithms are fair and reliable. Several strategies can help tackle bias in medical data before training algorithms [Chen et al. 2023; Wang, Chaudhari and Davatzikos 2023]. First, it is important to ensure that the dataset is representative of different ethnicities, genders, ages, and other relevant demographic factors. During data collection, a diverse

sampling strategy should be employed, sourcing data from varied populations to provide the algorithm with balanced training examples. This
diversity helps the algorithm generalize more effectively (see Chapter 1,
Section 1.1). Additionally, removing personal identifiers that could make
it easier for the algorithm to distinguish specific groups can help reduce
bias. Involving domain experts – such as doctors, nurses, and therapists –
throughout the data collection and bias detection process can further
improve the quality and fairness of the data. Addressing bias requires continuous effort and oversight to ensure that AI algorithms remain accurate,
fair, and equitable for all patients. This approach fosters the development
of ethically responsible AI technologies that uphold principles of equality
and social justice.

3.5 UNDERSTANDING HOW THE AI MAKES CHOICES (EXPLAINABILITY)

The problem of explainability refers to the ability to clearly understand
and describe how AI models make decisions and perform tasks. This issue
has become increasingly relevant with the rise of complex machine learning algorithms, such as deep neural networks, which can be challenging
to interpret, even for experts. In some nonmedical applications, explainability may not be essential. For example, in image or video recognition
systems using deep neural networks, the primary focus is accuracy rather
than understanding the model decision-making process. Similarly, in
video games where AI is used to optimize strategies – like Google AlphaGo
in the game Go – explainability is less important than the primary goal,
which is winning. In automatic translation software like Google Translate,
the AI uses complex models without explaining every decision as long as
the translation is accurate and comprehensible. Likewise, in streaming or
e-commerce services, AI algorithms provide personalized recommendations without explaining the process underlying why the software suggests
a product or content; the main goal is to match user preferences. In these
contexts, the effectiveness of the AI in completing the task is prioritized
over explaining its logic in detail. However, even in such cases, ensuring that the algorithms operate reliably and ethically remains crucial to
avoid potential discrimination or bias (see Section 3.4). AI models where
the decision-making process is opaque are often referred to as "black box
models".

In the medical field, understanding how and why an AI system makes
decisions, especially regarding diagnoses or treatment choices, is crucial.

For instance, if an AI predicts a high probability of a stroke, the patient will want the doctor to explain the "AI reasoning" behind this prediction. Consider a more specific scenario: A diabetic patient using an AI-based system for continuous blood glucose monitoring. The system analyzes real-time data to determine insulin doses needed to maintain safe blood sugar levels. One day, the AI recommends a significantly higher insulin dose than usual. This decision is based on a complex analysis of factors such as blood sugar levels, food intake, physical activity, and the patient's personal physiological variables. If the AI recommendation is not explainable, the physician and the patient may feel uncertain about the increased dose. Without understanding the AI reasoning behind the change, the doctor may hesitate to administer the higher dose due to the risk of hypoglycemia (low blood sugar). On the other hand, if the AI can explain that it recommended the higher insulin dose because of a rise in blood sugar from a carbohydrate-rich meal and lack of physical activity, the physician can better understand the rationale behind the decision and may feel more confident in following the recommendation. Similarly, the patient is more likely to accept the treatment change if they understand the reasoning behind the AI suggestion. In this way, explainability enhances trust in AI, enabling healthcare professionals and patients to feel more comfortable with AI decisions. It improves the management of the patient condition and reduces the risk of complications, as it ensures that dosing decisions are understandable and acceptable.

In Chapter 2, we discussed how AI can analyze large amounts of heterogeneous data, such as genetic tests, brain images, and clinical data, to identify patient characteristics and suggest personalized treatments. The explainability of AI is crucial in this process, as it helps doctors and patients understand how decisions are made based on the interaction of various data types. For example, consider a patient showing early signs of cognitive decline suspected of developing Alzheimer's disease. AI analyzes the patient's genetic data, clinical data, age, family history, and other risk factors to identify correlations that might be missed by traditional statistical methods. If the AI detects a combination of genetic variants and risk factors suggesting a high probability of Alzheimer's, it might recommend a personalized treatment plan. This plan could include lifestyle changes, specific pharmacological therapies, or participation in clinical trials exploring experimental treatments tailored to the patient's genetic profile. The explainability of AI is vital in this context, as it allows physicians to understand the rationale behind the AI recommendations. By showing which genetic variants and risk factors led to specific suggestions, AI provides transparency, enabling doctors to validate and clearly communicate the recommendations to the patient, ensuring active involvement in their personalized treatment plan.

Aside from trust and reliability, explainability is crucial for legal considerations and the question of liability discussed earlier. AI algorithms must be capable of providing counterfactual explanations – explaining what might have happened under different circumstances. If an algorithm recommends a specific treatment, it should be able to specify which alternatives were considered and why the recommended treatment is the best option. Without this transparency, the physician may hesitate to apply the AI-recommended therapy. If a patient suffers harm due to a decision made by the algorithm, and the doctor cannot explain the rationale behind the AI choice, the doctor could face legal liability. In such a case, the machine makes the decision, but the doctor is responsible, even if they cannot fully understand the algorithm's decision-making process. This aspect is problematic. These regulations governing medical practice are vital for ensuring patient safety, quality of care, and professional accountability. A lack of explainability in clinical decisions can hinder compliance with these rules.

Explainability in AI models is particularly valuable in medical research, especially when studying complex conditions like movement disorders. Parkinson's disease, one of the most recognized movement disorders, involves extensive data from diverse sources – clinical records, genetic sequences, and brain imaging. AI models can detect intricate patterns within these datasets, offering predictions or insights into disease progression. By applying explainability techniques, researchers can identify key factors – such as certain brain regions or combinations of clinical symptoms – that contribute most to the AI conclusions. For instance, they may discover that specific genetic markers or brain imaging features are strongly associated with Parkinson's onset or progression. This deeper understanding can lead to identifying early biomarkers for diagnosis and shedding light on the neurophysiological processes driving the disease. Such insights could also fuel the development of personalized treatments as researchers better understand which pathways to target, paving the way for more effective therapies tailored to the individual patient profile.

Researchers are developing new methodologies and algorithms to make AI models more transparent and interpretable. These advancements are essential to ensuring safe and effective AI use in medicine. Currently, several techniques could enhance AI explainability (see Box 3.2).

BOX 3.2   SOME TECHNIQUES TO MAKE AN AI EXPLAINABLE

Boosting results visualization. Choosing the correct visualization of AI results is critical in enhancing explainability, making the outputs more accessible to non-AI developers. For instance, consider an AI system analyzing stroke patient data to predict the risk of recurrence or monitor rehabilitation progress. Scatter plots[9] could illustrate relationships between critical variables like age, blood pressure, glucose levels, and the likelihood of recurrence. Histograms could compare patient characteristics across high-risk and low-risk groups. Dynamic time charts can track vital signs, such as blood pressure and heart rate, showing how they evolve during rehabilitation. These visual aids foster communication between the AI system and its users, ultimately promoting trust and informed decision-making.

Using implicitly explainable ML algorithms. Some ML algorithms, like decision trees, are known for their inherent explainability. A decision tree learns to make decisions by following a set of rules based on the features of the input data. It recursively divides the dataset into smaller subsets, each based on specific data attributes, until it arrives at a prediction or decision. This process creates an easy-to-interpret tree structure, making the decision-making process clear and understandable. For example, to develop a decision tree for diagnosing skin diseases, it is possible to gather a dataset containing relevant information, such as symptoms (itching, lesion size, color), patient characteristics, family history, and confirmed diagnoses. The decision tree then uses these inputs to learn rules associating symptoms with potential diagnoses. For example, when a patient presents with itching and a family history of eczema, the system may suggest eczema based on the rule: "If itching = yes and family history = yes, then diagnosis = eczema". This approach ensures transparency, allowing doctors to understand how the diagnosis was determined. Explainable machine learning algorithms, such as decision trees, linear regression, and logistic regression, enhance trust by offering interpretable results. Linear regression identifies the relationships between variables, while logistic regression is ideal for binary classification tasks like diagnosing a condition (yes/no), providing easily understood outputs for medical decision-making.

Features importance. Features importance is essential for understanding which patient characteristics have the most influence on AI predictions. For instance, in a model designed to predict heart disease risk based on factors like age, blood pressure, cholesterol levels, smoking status, and heart rate, there are methods to determine the relative significance of each factor in the AI decision-making process. One widely used technique is SHapley Additive exPlanations (SHAP), which assigns importance values to features by considering all possible combinations of them. For example, if the model predicts a high risk of heart disease for a particular patient, SHAP analysis might reveal that the patient age is the most significant factor, followed by blood pressure and cholesterol. This shows that age has a stronger impact on the prediction than other features. A graph illustrating SHAP values might highlight which features – such as high cholesterol – raise the risk significantly, while factors like a lower resting heart rate indicate better heart health. SHAP can also show how certain characteristics interact; for instance, it might reveal that blood pressure has a greater impact on predictions for patients over 60, while smoking has a larger influence on younger patients. By using SHAP, physicians gain a clearer understanding of how patient features drive model predictions, allowing for more personalized and informed decision-making in patient care.

A recent example of AI incorporating explainability techniques, as discussed in Box 3.2, is software developed through a collaboration between the Institute of Cognitive Science and Technology, the Institute of Biomedical Technology of the Italian National Research Council, the IRCCS Fondazione Santa Lucia, Policlinico Umberto I in Rome, the University of Pavia, and Sapienza University of Rome.[10] The goal of this system is to provide clear and accessible insights for the early diagnosis and monitoring of Alzheimer's disease progression.

NOTES

1. An investment bank is a financial institution that, unlike commercial banks, does not accept deposits. It offers financial advisory services, including helping companies manage their capital through mergers and acquisitions, underwriting securities, and facilitating investments. The primary role of an investment bank is to assist businesses and governments in raising capital by issuing stocks and bonds and offering strategic advice on financial transactions.
2. Through differential diagnosis, the doctors identify the most likely cause of a patient's symptoms. It involves a thorough evaluation of the patient's physical examination, diagnostic test results, and a comparison with a range of potential medical conditions. The goal is to systematically eliminate less likely conditions and narrow the options until the most probable diagnosis is determined. This process helps ensure patients receive the appropriate treatment for their specific condition.
3. https://glass.health/ai.
4. A clinical plan outlines the detailed strategies for treating and managing a patient's medical condition. It includes medication prescriptions, therapies, surgeries, specialist consultations, and other necessary interventions. The physician develops the plan based on the differential diagnosis, considering the patient's unique needs and circumstances.

5. https://nri.ntc.columbia.edu/.
6. https://neuralink.com/.
7. Transcranial magnetic stimulation (TMS) uses magnetic fields to stimulate specific brain areas. Transcranial direct current stimulation (tDCS) modulates neuronal activity by applying a mild electric current through electrodes placed on the scalp surface.
8. Cryptography is a technique used to enhance the security of information access. It involves converting data into an unreadable format, known as "encryption", only individuals with the correct "key" can decrypt and understand it. It creates a secret code that prevents unauthorized access to sensitive information.
9. A scatter plot illustrates the relationship between two variables by displaying data points corresponding to their value pairs on a coordinate plane. By examining the graph, it is possible to observe any patterns or correlations between the variables, such as whether they increase or decrease together or if there are any outliers or clusters that suggest a specific trend.
10. A detailed description of the project can be found on the following website: https://ctnlab.it/index.php/explain-medical-analysis-ema/

REFERENCES

Acemoglu, D. & Restrepo, P. [2020], *Robots and Jobs: Evidence from US Labor Markets*, Journal of Political Economy, 128, n. 6, pp. 2188–2244.

Bessen, J. E. [2016], *How Computer Automation Affects Occupations: Technology, Jobs, and Skills*, Boston Univ. School of Law, Law and Economics Research Paper, no. 15–49, doi: http://dx.doi.org/10.2139/ssrn.2690435.

Caligiore, D. [2022], *IA istruzioni per l'uso*, Bologna, Il Mulino.

Caligiore, D., Giocondo, F. & Silvetti, M. [2022], *The Neurodegenerative Elderly Syndrome (NES) Hypothesis: Alzheimer and Parkinson are Two Faces of the Same Disease*, IBRO Neuroscience Reports, 13, pp. 330–343.

Carrillo-Reid, L., Han, S., Yang, W., Akrouh, A. & Yuste, R. [2019], *Controlling Visually Guided Behavior by Holographic Recalling of Cortical Ensembles*, Cell, 178, n. 2, pp. 447–457.

Chen, Z., Zhang, J. M., Sarro, F. & Harman, M. [2023], *A Comprehensive Empirical Study of Bias Mitigation Methods for Machine Learning Classifiers*, ACM Transactions on Software Engineering and Methodology, 32, n. 4, pp. 1–30.

Coeckelbergh, M. [2016], *Responsibility and the Moral Phenomenology of Using Self-Driving Cars*, Applied Artificial Intelligence, 30, n. 8, pp. 748–757.

Ettinger, T., Berberian, M., Acosta, I., Cucca, A., Feigin, A., Genovese, D., … & Rizzo, J. R. [2023], *Art Therapy as a Comprehensive Complementary Treatment for Parkinson's Disease*, Frontiers in Human Neuroscience, 17, p. 1110531.

Frank, M. R., Autor, D., Bessen, J. E., Brynjolfsson, E., Cebrian, M., Deming, D. J., … & Rahwan, I. [2019], *Toward Understanding the Impact of Artificial Intelligence on Labor*, Proceedings of the National Academy of Sciences, 116, n. 14, pp. 6531–6539.

Freud, S. [1999], *The Interpretation of Dreams*, translated by Joyce Crick, Oxford, Oxford University Press.

Lauring, J. O., Ishizu, T., Kutlikova, H. H., Dörflinger, F., Haugbøl, S., Leder, H., ... & Pelowski, M. [2019], *Why Would Parkinson's Disease Lead to Sudden Changes in Creativity, Motivation, or Style with Visual Art?: A Review of Case Evidence and New Neurobiological, Contextual, and Genetic Hypotheses*, Neuroscience & Biobehavioral Reviews, 100, pp. 129–165.

Legrenzi, P. & Umiltà, C. [2018], *Molti inconsci per un cervello. Perché crediamo di sapere quello che non sappiamo*, Bologna, Il Mulino.

Liévin, V., Hother, C. E. & Winther, O. [2022], *Can Large Language Models Reason About Medical Questions?*, arXiv, 2207.08143, doi: https://doi.org/10.48550/arXiv.2207.08143.

Machado Sotomayor, M. J., Arufe-Giráldez, V., Ruíz-Rico, G. & Navarro-Patón, R. [2021], *Music Therapy and Parkinson's Disease: A Systematic Review from 2015–2020*, International Journal of Environmental Research and Public Health, 18, n. 21, p. 11618.

Mattioli, F., Maglianella, V., D'Antonio, S., Trimarco, E. & Caligiore, D. [2023], *Non-Invasive Brain Stimulation for Patients and Healthy Subjects: Current Challenges and Future Perspectives*, Journal of the Neurological Sciences, 456, p. 122825.

Nelson, J., McKinley, R. A., Phillips, C., McIntire, L., Goodyear, C., Kreiner, A. & Monforton, L. [2016], *The Effects of Transcranial Direct Current Stimulation (tDCS) on Multitasking Throughput Capacity*, Frontiers in Human Neuroscience, 10, 589, doi: https://doi.org/10.3389/fnhum.2016.00589.

Pagano, T. P., Loureiro, R. B., Lisboa, F. V., Peixoto, R. M., Guimarães, G. A., Cruz, G. O., ... & Nascimento, E. G. [2023], *Bias and Unfairness in Machine Learning Models: A Systematic Review on Datasets, Tools, Fairness Metrics, and Identification and Mitigation Methods*, Big Data and Cognitive Computing, 7, n. 1, p. 15.

Shen, G., Horikawa, T., Majima, K. & Kamitani, Y. [2017], *Deep Image Reconstruction from Human Brain Activity*, bioRxiv, 240317, doi: https://doi.org/10.1101/240317.

Wang, G., Badal, A., Jia, X., Maltz, J. S., Mueller, K., Myers, K. J., ... & Zeng, R. [2022], *Development of Metaverse for Intelligent Healthcare*, Nature Machine Intelligence, 4, n. 11, pp. 922–929.

Wang, R., Chaudhari, P. & Davatzikos, C. [2023], *Bias in Machine Learning Models Can be Significantly Mitigated by Careful Training: Evidence from Neuroimaging Studies*, Proceedings of the National Academy of Sciences, 120, n. 6, p. e2211613120.

Yuste, R., Goering, S., Bi, G., Carmena, J. M., Carter, A., Fins, J. J., ... & Wolpaw, J. [2017], *Four Ethical Priorities for Neurotechnologies and AI*, Nature News, 551, n. 7679, pp. 159–163.

CHAPTER 4

The Medicine of the Future

A Contamination of Skills and Disciplines

4.1 TREATING THE DISEASE INDIVIDUALLY BUT LOOKING AT THE 'SYSTEM'

A groundbreaking aspect of artificial intelligence (AI) in medicine is its ability to address diseases by combining personalized attention to individual traits – such as genetic factors and lifestyle – with a holistic understanding of the person as a whole system, considering the interactions between the brain, body, and environment. Genetics plays a crucial role in personalized treatments, as genes act like small pieces of a puzzle, determining how our body develops and functions. They influence an individual susceptibility to certain diseases and responses to medications. Genetic analysis can identify mutations and help select the most appropriate therapies for each patient. However, genetics is not the only factor. A person's lifestyle and environment can significantly affect their health and treatment outcomes. For instance, people living in highly polluted or noisy environments may be more prone to certain health conditions, such as respiratory problems or heightened stress levels.

Lifestyle and environmental factors can directly influence genetics, for example, by affecting *telomeres* – DNA structures that protect chromosomes

DOI: 10.1201/9781003606130-4

and help preserve the integrity of a cell genetic material. Cells in the body continually replicate to replace those that have reached the end of their life cycle. During this process, telomeres gradually shorten. Once they become too short, the cell can no longer reproduce and dies. This phenomenon is associated with aging and the onset of diseases. Although telomere shortening is a natural process, recent research indicates that lifestyle choices – such as body mass index, smoking, physical activity, stress levels, and diet – can influence the rate of telomere shortening. A healthier lifestyle may help slow down this process, delaying aging-related conditions. Studies by Epel et al. (2004) and Lumera and De Vivo (2020) support this connection, suggesting that how we live can significantly affect our DNA and overall health.

ML algorithms can explore the interaction between genetic factors and lifestyle aspects. As discussed in previous chapters, these algorithms can *integrate and analyze diverse data from various domains*, such as genetic, physiological, and environmental information, to predict the likelihood of disease onset and progression. They can also help tailor treatments to individual patients and enhance therapy effectiveness. By leveraging AI in this way, healthcare professionals can approach disease treatment more holistically, considering not only the specific characteristics of the patient but also the entire brain-body-environment system. This aspect allows for effective personalized medical care.

Several international research projects use AI to explore the interaction between lifestyle and health. One example is the GATEKEEPER project,[1] which involves countries like Spain, Germany, the United Kingdom, Poland, Cyprus, Greece, and Italy. One of its goals is to assess whether AI can encourage people to adopt healthier lifestyles. The project investigates how smartphone AI-powered wellness apps can influence users behaviors by providing personalized suggestions based on integrated data, such as physical activity levels, diet, and other health-related factors. GATEKEEPER promotes the digital transformation of the European healthcare sector by developing an open, scalable platform that leverages advanced technologies like AI, the Internet of Things (IoT), and blockchain. This platform securely collects and manages patient health data while integrating it with environmental and social information, enabling more holistic and data-driven healthcare solutions.

Studying a disease through a systems approach involves more than considering the patient environment. It also requires analyzing both the role of brain regions or other bodily systems directly linked to the disease and the potential effects of interactions between these areas and other parts of the body that are not traditionally associated with the illness. This holistic view fosters interdisciplinary and multi-methodological discussions, leading to innovative solutions that might not emerge from approaches that focus on isolated elements, such as a single brain region, gene, or environmental factor. By broadening the scope to include how various systems within the body interact, researchers can generate new hypotheses and treatment strategies that offer a more comprehensive understanding of disease mechanisms. This aspect contrasts with more reductionist approaches focusing on a single specific component in isolation. Such broader perspectives are key to advancing medical science.

The systems approach can also help generate new hypotheses about the relationships between different diseases. An example comes from research at the Institute of Cognitive Sciences and Technologies of the CNR, which proposes the Neurodegenerative Elderly Syndrome (NES). According to this hypothesis, Alzheimer's and Parkinson's – two of the most common neurodegenerative diseases – might represent different aspects of the same condition, NES [Caligiore, Giocondo and Silvetti 2022]. Traditionally, Alzheimer's and Parkinson's are treated as distinct diseases. However, recent scientific research suggests significant overlaps in their causal factors, mechanisms of onset, progression, symptoms, and the brain regions affected. For example, Parkinson's, primarily viewed as a movement disorder, often involves cognitive and emotional impairments (such as memory issues, apathy, or depression), which are also common in Alzheimer's. Conversely, Alzheimer's, typically known for cognitive decline, may present with motor issues similar to Parkinson's disease. These shared features suggest that the two diseases may not be entirely separate but interconnected manifestations of a broader neurodegenerative process.

According to the NES hypothesis, Alzheimer's and Parkinson's disease may arise from the same underlying neurodegenerative processes, which begin years before obvious symptoms appear. These mechanisms spread gradually, becoming detectable only later in life. This hypothesis is based on an interdisciplinary and systemic analysis of scientific literature across various fields, such as genetics and neurophysiology. NES proposes that both diseases follow a three-stage progression (see Figure 4.1). In the early NES stage, which begins long before clinical symptoms emerge, there is a gradual loss of neurons responsible for producing two key neuromodulators: noradrenaline and serotonin. This neuron loss likely starts in the locus coeruleus (LC), which produces noradrenaline, and the dorsal raphe

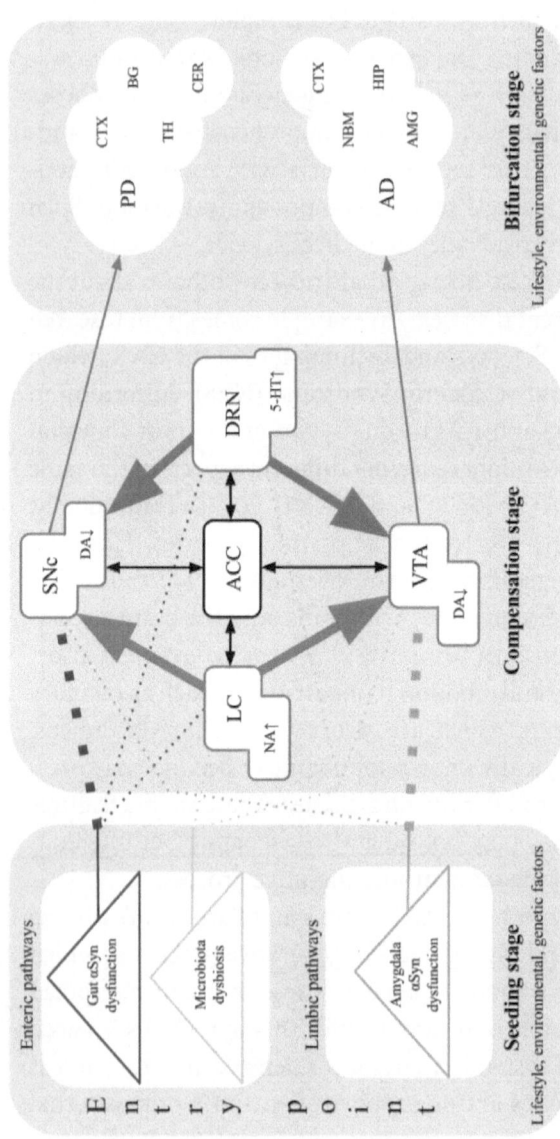

FIGURE 4.1 The progression of the three NES stages. During the seeding stage (left) the different types of seed could set different initial pathways (dashed lines) towards a possible future development of NES in AD or PD. The different dashed line thickness indicates the different initial probability that NES could become AD or PD (large thickness, greater probability). The initial neurodegenerative trajectory is influenced by lifestyle, genetics, and environmental factors (bifurcation stage), which make the seeding stage determine only probabilistically the future outcome of the bifurcation stage leading to AD or PD. In the compensation stage (middle) the ACC could upregulate the LC and/or DRN activity to recover the DA loss in SNc or VTA (thicker arrows), according to a cost-benefit trade-off. In the bifurcation stage (right) NES becomes AD or PD. Lifestyle, environmental, and genetic factors could affect both the seeding and the bifurcation stages. Abbreviations: AD: Alzheimer's disease; α Syn: alpha-synuclein; AMG: amygdala; BG: basal ganglia; 5-HT: serotonin; CER: cerebellum; CTX: cortex; DA: dopamine; DRN: dorsal raphe nucleus; HIP: hippocampus; LC: locus coeruleus; ACC: anterior cingulate cortex; NA: noradrenaline; NBM: nucleus basalis of Meynert; SNc: substantia nigra pars compacta; PD: Parkinson's disease; TH: thalamus; VTA: ventral tegmental area.

Source: Caligiore, Giocondo and Silvetti [2022].

nucleus (DRN), responsible for serotonin. This initial damage results from a dysfunction of alpha-synuclein (αSyn), a protein found throughout the body. Although this early-stage neurodegeneration does not produce recognizable symptoms of Alzheimer's or Parkinson's, it is influenced by various genetic, environmental, or lifestyle factors, termed "seeds". Alpha-synuclein dysfunction may begin in different parts of the body, such as the brain or the gut, and can travel through pathways like the gut-brain axis. These early "seeds" determine how the disease will progress and whether it will develop into Parkinson's or Alzheimer's. This stage is referred to as the "seeding stage".

In the second phase of NES, dysfunction affects neurons that produce dopamine. These neurons are in two key regions of the brain: the ventral tegmental area (important for managing cognitive and motivational functions) and the substantia nigra pars compacta (critical for motor control). Despite this dysfunction, the typical clinical symptoms of Alzheimer's and Parkinson's disease do not yet manifest. This is due to compensatory mechanisms that help maintain the balance of neuromodulator levels in the brain. We refer to this stage as the "compensation stage".

Finally, we arrive at the third stage of NES, which we refer to as the bifurcation stage. Norepinephrine and serotonin can no longer compensate for dopamine dysfunction. NES then diverges into Alzheimer's disease if the ventral tegmental area is most affected or Parkinson's disease if the substantia nigra pars compacta is primarily involved. The eventual outcome – whether Alzheimer's or Parkinson's – depends on genetic, environmental, and lifestyle factors, which can either reinforce or modify the neurodegenerative trajectory that began in the seeding phase.

If confirmed by future empirical studies, the NES hypothesis, emerging from a systemic and multidisciplinary approach, could transform research in the field of these two neurodegenerative diseases. It would open new possibilities for early diagnosis and the development of therapies to be implemented at very early stages – before explicit clinical symptoms manifest – thus enabling a much more effective counteraction to neurodegenerative processes. Currently, we are developing machine learning algorithms to integrate and analyze large amounts of heterogeneous data (clinical, genetic, MRI) from Alzheimer's and Parkinson's, provided by international scientific research databases such as ADNI and PPMI.[2] We are going to use such algorithms to test the NES hypothesis. Specifically, the aim is to identify shared neurodegeneration trajectories between the two diseases.

To treat diseases in a person-centered yet systemic manner, it is essential to consider the interaction of diverse data related to the brain-body-environment system and to provide therapies tailored to the individual patient's needs. This approach necessitates close interdisciplinary collaboration between physicians and experts from various fields, as well as stronger patient involvement. The goal is to develop innovative and targeted treatments that address the patient's unique characteristics and environmental factors. The following section delves into the effects of this interdisciplinary collaboration.

4.2 A NEW RENAISSANCE IN CONTAMINATION OF SKILLS AND DISCIPLINES

The rise of AI encourages cross-disciplinary collaboration, which will significantly impact the future of medical research. Developing an AI model capable of analyzing a diverse range of data related to the brain, body, and environment – or replicating the function of various brain and body systems (digital twins) – requires expertise from multiple fields such as neuroscience, biology, computer science, engineering, mathematics, psychology, and AI. Over time, researchers building these models accumulate knowledge across these disciplines, gaining a deeper understanding of some areas while becoming less involved in others. They learn to select the most relevant computational inputs for the mathematical equations that define the model. However, it is essential to collaborate with specialists in specific fields to refine and adapt the model to address the problem at hand. For instance, to create an AI model that studies the mechanisms behind Alzheimer's disease, it is crucial to involve neurobiologists, neuroscientists, or clinicians who specialize in understanding the disease pathogenesis. Experts in these fields are critical for interpreting the model results within the context of the disease. In addition, they could help to select and process the data used in model training to minimize biases (as discussed in Chapter 3, Section 3.4).

The AI model fosters multidisciplinary collaboration by bringing together diverse perspectives and expertise. Researchers from different fields can share insights, leading to a deeper understanding of disease mechanisms and innovative approaches in medical research. As seen in previous chapters, collaboration between physicians, geneticists, biologists, and data scientists can uncover new opportunities for disease prevention, early diagnosis, risk factor analysis, and the development of novel treatments targeting the brain-body-environment system. Additionally, *cross-disciplinary partnerships*, such as those between medicine and the human sciences (e.g., psychology and social sciences), can yield equally valuable results. Sociology can provide insights into how social factors impact health, while psychology helps us understand the connection between mental well-being and physical health. These disciplines, combined with AI, can help simulate and analyze the effects of various factors on health. For example, consider studying the impact of major depression, a neuropsychological illness that affects both individual and social well-being. We could collect diverse data from various sources, such as interviews, questionnaires, demographic information (age, gender, education, etc.), MRI scans, and social factors like employment or marital status. AI can process this heterogeneous data to identify patterns and make classifications. A machine learning model could be trained to differentiate between depressed and nondepressed individuals based on factors such as occupation, social activities, and environmental influences. This would help identify which factors are most relevant for diagnosing depression, as seen in the feature importance analysis. For example, if "occupation" is found to be a significant factor, this would indicate that employment status plays a major role in diagnosing depression. Conversely, if "age" shows low importance, it suggests age has minimal impact on the likelihood of depression. AI can also be used to predict how depression might influence broader social factors, like unemployment or crime rates, shedding light on the societal impact of mental health. Additionally, machine learning could reveal correlations between depression and demographic or social data, potentially identifying subtypes of depression or groups at higher risk (cluster analysis). By leveraging AI in this way, we can gain a more comprehensive understanding of both individual and societal factors that contribute to mental health conditions.

In genetics, integrating biology, computer science, and mathematics has led to the development of advanced DNA sequencing technologies[3] capable of analyzing human genomes in just a few hours. This breakthrough has enabled rapid identification of gene mutations and other genetic variations linked to diseases. High-throughput sequencing, however, generates vast amounts of data that are difficult to manage with traditional statistical methods. AI plays a crucial role in automating and enhancing this process. For example, deep learning algorithms can process large-scale DNA sequencing datasets to detect genetic variants, mutations, and other features that could affect individual health [Alharbi and Rashid 2022]. AI can also integrate genomic data (such as DNA sequences, gene expression, and proteomics) to provide critical insights into the genetic foundations of diseases and their progression.

AI could enhance the efficiency of *gene therapy*, a promising treatment aimed at correcting or preventing genetic diseases by manipulating a patient's genes. It can involve introducing a healthy gene into cells with a defective or missing gene or altering DNA segments. Gene therapy is a multidisciplinary field, merging genetics, molecular biology, biotechnology, medicine, and pharmacology. Genetics provides insights into gene and cell structure and the causes of genetic diseases. Molecular biology manipulates DNA, while biotechnology enables large-scale production of genetic material; medicine brings knowledge of diseases and their mechanisms. Pharmacology contributes methods to develop gene-targeting drugs. AI can play a crucial role in gene therapy by accurately identifying genes for modification, refining therapy designs, and detecting potential side effects [Vilhekar and Rawekar 2024]. AI can analyze large genomic datasets to identify correlations between specific genes and diseases (e.g., which genes are overexpressed in certain pathological conditions), guiding the selection of genes to modify for therapeutic purposes. Machine learning algorithms can predict how genetic mutations will impact genes, helping assess whether modifications will lead to a disease or improve a condition. Additionally, AI-powered models, such as digital twins, can simulate gene and protein behavior within an organism, enabling the assessment of potential effects before making actual genetic changes.

AI will significantly impact synthetic biology, an interdisciplinary field that combines biology, bioengineering, computer science, and mathematics to create artificial organisms and redesign existing ones. This field holds great potential for developing new treatments for genetic diseases and creating organisms capable of producing useful substances, such as novel drugs. AI can analyze large amounts of genomic and biological data, uncovering hidden relationships that help researchers better understand molecular biology and explore new pathways for synthetic organisms. For example, AI could support the design of synthetic bacteria capable of producing human insulin in sufficient quantities to treat diabetes. By analyzing large genomic datasets, AI could predict genetic modifications that could optimize insulin production in bacteria. Additionally, digital twins of synthetic bacteria could simulate their behavior in various environments, helping researchers predict and optimize conditions for insulin production. Similarly, AI models could aid in the design of synthetic viruses modified to carry therapeutic genes for treating genetic diseases. AI could also assist in designing custom synthetic organs for transplantation, thus reducing reliance on human donors. These AI-driven advancements hold promise for transforming medicine and biotechnology by enhancing precision and efficiency in synthetic biology.

The convergence of biomedical engineering, materials science, and biology has led to the development of advanced medical devices and cutting-edge diagnostic and therapeutic techniques. Biocompatible materials, such as bioactive glass, have played a key role in this progress. Bioactive glass, composed of silicates and other materials, interacts with biological tissues, forming stable chemical bonds that promote cell growth. This property makes it ideal for applications in joint prostheses, dental implants, and ocular implants, as it facilitates tissue regeneration and enhances the durability of medical devices. AI can significantly accelerate the synthesis of new biocompatible materials by analyzing large databases containing material properties and suggesting optimal component combinations. Machine learning algorithms, for example, can streamline the identification of materials with antimicrobial properties, which are crucial for producing infection-resistant surfaces in implantable devices like pacemakers. Using predictive models, such as regression algorithms or digital twins of materials, AI can estimate the antimicrobial properties of new materials before they are synthesized or tested experimentally. This helps focus research on the most promising candidates, speeding up the development process. Moreover, AI can optimize the production processes of biomaterials, improving efficiency and reducing costs. For instance, regressors can predict when production equipment needs maintenance, minimizing downtime and maximizing efficiency. AI can also monitor production processes in real-time to detect defects or anomalies in materials, ensuring that high-quality standards are met. Automated quality control powered by AI helps minimize human errors and ensures that the biomaterials produced are of the highest quality.

The intersection of even the most disparate disciplines can offer innovative perspectives in understanding complex phenomena. For example, a few years ago, an exhibition called *As Above, As Below* explored the striking similarities between the far reaches of the universe and the neural networks within the human brain. Using digital projections, virtual reality, and interactive multimedia, an interdisciplinary team of artists, astrophysicists, and neuroscientists created six pieces of art that highlighted parallels between intergalactic networks and neural networks [Neyrinck et al. 2020]. Similarly, a recent study led by astrophysicist Franco Vazza of the University of Bologna and neurosurgeon Alberto Feletti of the University of Verona revealed fascinating *similarities between the human brain and the universe*. The research, published in Frontiers in Physics, employed methods from cosmology, neuroscience, and network analysis to examine the structural and morphological properties of both the brain network of neurons and the cosmic network of galaxies (Figure 4.2). Despite the vastly different physical interactions in these two networks, microscopic and telescopic observations revealed striking parallels. The study suggested that the self-organization of both systems may be governed by *similar principles* of network dynamics [Vazza and Feletti 2020]. Both networks consist of well-defined structures, with approximately 10^{10}–10^{11} nodes connected by filaments – neurons for the brain and galaxies for the universe. Interestingly, both neurons and galaxies have a scale radius[4] that is a tiny fraction of the length of the filaments to which they are connected. The brain is composed of water (77–78%), lipids (10–12%), proteins (8%), carbohydrates (1%), soluble organic substances (2%), and salt (1%), while the universe comprises dark energy (73%), dark matter (22.5%), ordinary matter (4.4%), and photons and neutrinos (≤0.1%). Remarkably, in both systems, around 75% of the mass/energy is made up of seemingly passive material – water in the brain and dark energy[5] in the universe – that plays an indirect role in their internal structures. This research underscores how combining different fields of study can illuminate surprising connections between diverse systems, offering new insights into the fundamental nature of both the brain and the cosmos [Pfeifer et al. 2020].

Studying the similarities between the universe and the brain can foster interdisciplinary collaborations between astronomers, physicists, neuroscientists, and physicians, leading to novel approaches in brain disease research that leverage knowledge and tools from various fields. This collaboration can drive the development of advanced imaging, monitoring, and data analysis techniques that enhance our ability to study the brain in greater detail, improving our understanding of neurological disorders. For example, advanced imaging technologies inspired by astronomical telescopes could significantly improve medical imaging devices used to study the brain. Astronomy often employs noninvasive imaging techniques to observe distant celestial objects with high spatial resolution. Translating this "far-field spatial resolution" capability into medical imaging could lead to noninvasive methods for visualizing brain activity, reducing the

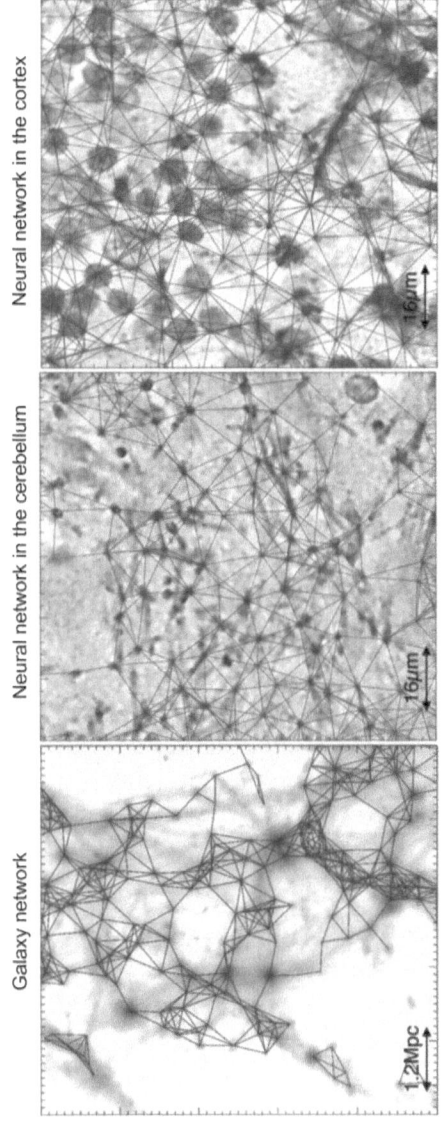

FIGURE 4.2 Computer simulation depicting the distribution of the neuronal network in two areas of the human cerebellum (center and right: cerebellum and cortex, respectively). "Mpc" stands for megaparsec (1 Mpc = 3.085 × 10²⁴ cm), a unit used in cosmology to measure vast distances in the Universe. "μm" stands for micron, a subunit of the centimeter, equal to one thousandth of a centimeter (1 μm = 1 × 10⁻⁴ cm).

Source: Adapted from *Vazza* and Feletti [2020].

need for invasive procedures and lowering risks for patients. Furthermore, this could provide highly detailed views of brain structures at microscopic levels, aiding in the early diagnosis of brain diseases such as Alzheimer's or Parkinson's. This cross-disciplinary approach holds the potential to revolutionize brain research by applying the sophisticated imaging techniques developed for space exploration to neuroscience, paving the way for innovative methods of diagnosis and treatment.

Both star systems and the brain are characterized by complex dynamic balances and interconnected networks, where the interactions between various components are essential for maintaining stability and supporting specific functions. This analogy can enhance our understanding of the complexity of both systems. In a stellar system, a dynamic equilibrium exists between gravitational forces that aim to collapse the star and the internal pressure created by nuclear reactions that push outward. This balance is crucial for the long-term stability of the star. Similarly, the brain operates within a dynamic equilibrium between excitatory and inhibitory activities of nerve cells. Neural homeostasis is vital for maintaining a balanced neural environment, which is essential for cognitive functions like memory, attention, and problem-solving. The parallels between the balance of a star system and the balance of cognitive functions in the brain could offer valuable insights into the study of neurological and psychiatric diseases. Algorithms inspired by the study of cosmic networks and their dynamic equilibrium could potentially be applied to analyze neuronal interconnections in the human brain.

This comparative approach could also be applied to brain waves research, as many psychiatric and neurological disorders are associated with alterations in these waves. By drawing inspiration from how astronomers study cosmic waves[6] from deep space, the analysis of brain waves could be enhanced. The study of cosmic waves has required the development of highly sensitive detection techniques, such as laser interferometers. These advancements in sensing technologies could influence the creation of more precise and noninvasive tools for monitoring brain activity. Additionally, in cosmic wave analysis, researchers must isolate weak signals from a background of noise. Developing techniques to differentiate weak signals from noise could similarly improve the detection and interpretation of low-intensity neural signals in brainwave analysis.

The examples discussed in this section highlight how future cross-disciplinary collaboration could address complex medical challenges and lead to more effective solutions. Interdisciplinary partnerships can significantly advance medical research by fostering innovation and facilitating the translation of scientific discoveries into clinical applications (*translational research*). By *bridging the gap between basic science and clinical practice*, these collaborations can help develop new, person-centered approaches to therapy and prevention. Ultimately, such collaboration will enable the creation of more holistic, targeted medical treatments that reflect the latest advancements in multiple fields.

4.3 STUDYING A DISEASE WITHOUT GUINEA PIGS

We have explored how AI can analyze large, heterogeneous datasets – such as biological, psychological, or genetic data – to identify correlations between risk factors and the onset of diseases or to develop predictive models that help researchers and physicians understand the *systemic mechanisms* behind disease progression. By combining AI with insights from computational mechanics, researchers can create digital twins of biological systems. These virtual replicas allow simulations of the effects of a drug or interactions between multiple drugs on a living organism. Using digital twins, researchers can study the underlying physiological processes of diseases through computer simulations, thus reducing the need for animal testing in drug development (see Chapter 2, Section 2.2).

Before therapies identified through AI can be applied clinically, they must still undergo additional experimental validation on animals and humans, meaning the complete replacement of guinea pigs in medical research is not yet feasible. Currently, AI can complement medical research by reducing the number of animal and human trials, but it cannot eliminate the need for these experiments. When might the use of guinea pigs be avoided? The main challenges lie in the complexity of the human organism – its interactions with the environment and the experiences accumulated over a lifetime – as well as the significant differences between humans and the animal species typically used in research. Two approaches offer potential solutions. The bottom-up approach suggests that for AI to replace animal models, it will need to generate digital twins that are far more detailed than those available today. AI simulates the human organism's function, modeling interactions between cells, tissues, organs, and systems. Achieving this requires extensive biological, medical, and physiological knowledge, alongside vast amounts of data. Moreover, AI must predict how an organism will react to treatments or therapies, ensuring safety without harming virtual patients. The top-down approach, by contrast, focuses less on replicating the organism's every detail. Instead, it seeks to capture the essence of the organism-environment system through AI and mathematical models, emphasizing key interactions rather than exhaustive replication. Both approaches aim to eventually minimize and perhaps eliminate the use of animal models in scientific research.

It is difficult to say which of the two approaches, bottom-up or top-down, is better, as each has advantages and limitations. The bottom-up approach, which focuses on reproducing detailed aspects of an organism, offers high specificity and precision. However, it comes with challenges, such as high computational costs, requiring computing power and storage for handling vast amounts of heterogeneous data. Even with future supercomputers, modeling something as complex as the human body in extreme detail could make analysis overwhelming and difficult to interpret. In contrast, the top-down approach emphasizes the broader system rather than intricate details, reducing the computing resources needed and making the model easier to understand and analyze. However, the main challenge lies in deciding what to simplify or omit. In this method, determining which elements to prioritize for reproduction is a critical decision that could affect the model accuracy and usefulness. Ultimately, the choice between the two approaches depends on the task. A hybrid approach, combining elements of both, may be the best way forward. For example, a highly detailed bottom-up model might be used to simulate critical parts of the system, such as the retina when studying vision diseases, while a simplified top-down model could represent other interconnected systems. This approach balances precision with computational efficiency, harnessing the strengths of both methodologies.

Regardless of the approach, establishing strict standards for evaluating AI simulations remains essential to ensure the accuracy and reproducibility of the results. Using AI to replace animal models in scientific research will require even stronger collaboration across disciplines – between scientists, software engineers, and experts in ethics and regulation. The synergies between AI and other emerging technologies also have the potential to reduce reliance on animal testing. For example, organ-on-a-chip devices, which simulate the functions of human tissues and organs, could allow researchers to test drugs and chemical compounds before moving to animal studies. Significant advancements in this area suggest that combining AI and these technologies may offer a more ethical and effective alternative to animal testing in future research.

4.4 TRAINING DOCTORS AND PATIENTS
FOR MINDFUL USE OF AI

Throughout this book, we have emphasized the importance of understanding how AI algorithms work to use them effectively and responsibly while acknowledging their limitations. AI could positively transform medicine, but its safe and competent application requires thorough training for healthcare professionals and patients. This aspect ensures that AI is applied effectively, reducing errors, safeguarding patients and caregivers, and maximizing its potential in healthcare.

Adequate training enables healthcare workers to effectively use AI tools, ensuring they *can interpret* and apply *AI results* appropriately in clinical practice. This training should combine technical and clinical knowledge. Healthcare professionals, even if not directly involved in algorithm development, must understand the fundamentals of AI and its application in medicine, including how machine learning algorithms work and how they analyze medical data. Several examples of AI applications in healthcare should be provided, demonstrating how AI improves diagnosis, prognosis, and treatment. As discussed throughout this book, AI can have limitations, such as errors resulting from incomplete or inaccurate training data. Doctors must be aware of these limitations and develop the ability to critically assess AI-generated data. To ensure safe and effective use, AI training should be integrated into medical university programs and ongoing professional development courses for healthcare professionals.

In this book, we have emphasized that while AI can provide insights based on medical data analysis, it cannot offer empathy or emotional support. These human qualities enable healthcare professionals to consider also personal and environmental factors affecting a patient's health, aside from medical data. Humans also possess the ability to offer empathy, understanding, and emotional support – critical components for patient well-being and fostering a trust-based doctor-patient relationship. Therefore, medical training programs should focus on developing technical and emotional intelligence. They should help healthcare professionals cultivate soft skills, such as empathy and emotional understanding, which complement AI technology. Additionally, courses should address aspects where AI falls short – like the caring relationship and emotional connection, which are vital for comprehensive patient care [Chell and Athayde 2011; Heckman and Kautz 2012]. These human qualities are essential across all professional environments, ensuring a balance between technological advancement and *the human touch*.

Patients also need to be educated on the mindful use of the new tools offered by AI, helping them better understand its benefits and limitations. The focus of patient education should not be on the technical aspects but on what AI can do, its boundaries, and how it could enhance awareness of one health, such as through self-monitoring that empowers patients to take a more active role in their care. Education should be clear, simple, and free from technical jargon, using accessible formats like informative articles, books, and videos. These materials should include concrete examples of AI applications in the medical field, such as prevention, early diagnosis, and the development of novel therapies. AI can be introduced to patients as a virtual assistant that aids doctors in making quicker and more accurate data interpretations. For example, in cases of hypothyroidism symptoms, AI can help confirm a doctor's diagnosis by swiftly analyzing relevant data, enabling timely interventions. However, it is important for patients to understand that AI is a supportive tool, not a replacement for the experience and expertise of healthcare professionals. While AI provides valuable insights, the final decision on diagnosis and treatment must always rest with a qualified doctor, who will integrate AI-generated data and their clinical judgment. This aspect ensures patients recognize that AI, while useful, is only one element in the broader context of medical care. Additionally, patients must be informed about the potential risks to their privacy and security when sharing personal data and be encouraged to adopt cautious behaviors in managing their information (see Chapter 3, Section 3.4).

4.5 A HIGH QUALITY OF CARE DESPITE THE GROWING POPULATION

The global population now exceeds 7.8 billion, and given the rapid growth in recent decades – an increase of roughly 1.24 billion people since 2000 according to the WHO – this number is expected to rise. This growth will likely increase demand for healthcare services, which could strain resources. If healthcare systems are not equipped to meet the growing demand, the quality of care may suffer. For instance, there could be shortages of doctors, hospital beds, and medications. Furthermore, this growing population could deepen inequalities in healthcare access, as people in remote or impoverished areas may face more barriers to quality care than those in wealthier regions. Other challenges linked to population growth include the spread of infectious diseases and pollution. Higher population density, especially in urban areas, raises the risk of disease outbreaks, while increased industrial activity and vehicle use contribute to air, noise, and water pollution, all of which negatively impact public health. Additionally, a larger population will demand more essential resources like food, water, and energy. If these resources become scarce, conflicts may arise, further jeopardizing people's health. It is crucial to address these factors now to ensure that healthcare remains accessible and of high quality for everyone in the future.

Can AI help maintain high-quality medical care despite population growth? The answer is "yes". AI can optimize healthcare resource management, improving access to growing patient numbers. For example, an AI system could use regression analysis to forecast healthcare demand and streamline hospital logistics. In this way, AI could predict the necessary storage for medical supplies, manage inventory, and organize medical transport efficiently. Below is an overview of the steps involved in developing and training an AI model for this purpose:

1. **Data collection**: Gather historical data on hospital operations, including medical supply usage, delivery schedules, and emergency transport details.

2. **Feature selection**: Identify key factors, such as patient volume, supply requirements, and warehouse capacity, to accurately predict future logistical needs.

3. **Model creation**: Develop a machine learning model based on regression analysis to connect these variables with desired outcomes, such as optimal medical resource distribution.

4. **Training**: Train the AI model using historical data, allowing it to learn the relationships between the variables and their outcomes, like predicting increased demand for supplies as patient numbers rise.

5. **Validation**: Test the model with new, independent data to evaluate its accuracy in making predictions for unforeseen situations, such as during medical emergency surges.

6. **Optimization**: Refine the model by retraining it with additional data or tweaking parameters to improve its precision in predicting logistical needs.

By effectively predicting future needs, AI can help healthcare systems anticipate and address surges in patient demand, ensuring that medical supplies, space, and transportation are available when needed, thus preserving the quality of care in a growing population. Once the AI model has demonstrated accuracy and reliability, it can actively assist healthcare professionals in optimizing hospital logistics. For example, the system could help ensure the efficient distribution of medical resources during periods of high demand, maintaining a steady, timely flow of services. This support enables healthcare providers to efficiently manage patient surges, ensuring optimal allocation of resources like supplies, beds, and staff to meet the needs of a growing population.

Regression models like the one described above can analyze large amounts of historical data, such as disease trends, surgeries, and demographic information, to identify patterns. For example, the model can predict seasonal disease peaks or increased doctor visits among specific age groups. With these predictions, AI can help plan for necessary resources like hospital beds, medical personnel, supplies, and equipment, preventing overuse and underuse of facilities. For instance, during a sudden flu outbreak in a densely populated area, AI could monitor real-time epidemiological data, patient demands, and healthcare resources. By analyzing historical data and current availability, AI can predict peak demands across hospitals. If a spike in flu cases is detected in a specific neighborhood, AI can assess the availability of medical staff, supplies, and equipment, forecasting potential shortages and suggesting optimal resource allocation. For example, AI might recommend transferring doctors, nurses, and medical supplies to hospitals with higher demand.

Machine learning algorithms can analyze healthcare personnel work patterns and predict peak activity periods, enabling more efficient shift planning. This feature ensures adequate coverage, with enough doctors and nurses available during high-demand times. AI systems can also monitor hospital bed occupancy, medical equipment use, and drug stock levels in real-time. If a hospital approaches maximum capacity, AI can issue early warnings to healthcare managers, allowing them to take timely actions, such as transferring patients to hospitals with available resources. This proactive approach can help maintain care quality and prevent system overloads.

In previous chapters, we have highlighted several ways AI can support physicians in administrative tasks and the analysis of diverse data, such as clinical test results, diagnostic images, and genetic information. AI can enable faster and accurate diagnoses, and the development of personalized treatment plans. By assisting with these tasks, AI helps physicians manage their time more effectively with patients, which improves care quality, especially when patient numbers are rising. Additionally, as discussed in Chapter 1, Section 1.3, telemedicine allows AI to monitor patients remotely in real-time. This enables physicians to receive prompt alerts about changes in a patient's vital signs, facilitating timely and precise interventions. This remote monitoring helps physicians maintain high-quality care, even as the population grows, by allowing them to oversee the conditions of multiple patients and intervene when necessary.

The digital patient implementation (see Chapter 2, para. 2), which simulates the effects of computer-assisted therapies, can play a crucial role in maintaining the quality of care despite the growing global population. The digital patient can be tailored to individual needs, factoring in genetic predispositions, lifestyle, and responses to treatment. This level of personalization is difficult to achieve with traditional medical tools, particularly as the population expands. Personalized care improves treatment effectiveness while minimizing side effects. AI allows the digital patient to speed up drug and therapy development. This aspect is crucial for providing high-quality care to a growing population. By simulating treatment responses, AI enhances research efficiency and precision, enabling faster innovation and better-targeted treatments.

NOTES

1. You can find a description of the project on the following website: https://www.gatekeeper-project.eu/.
2. Alzheimer's Disease Neuroimaging Initiative (ADNI): https://adni.loni.usc.edu/; PPMI (Parkinson's Progression Markers Initiative): https://www.ppmi-info.org/.
3. DNA sequencing is the process of determining the precise order of the four nitrogenous bases – adenine (A), thymine (T), cytosine (C), and guanine (G) – that make up a DNA molecule. This unique sequence of bases forms the genetic code, which dictates the traits, functions, and biological processes of an organism.
4. The scale radius refers to a characteristic length or specific scale that describes the distribution of structures related to the phenomena being studied, such as galaxy clusters in astronomy or neurons in the brain.
5. Dark energy is a hypothetical form of energy thought to permeate all empty space in the universe. It acts as a force that accelerates the universe expansion over time, counteracting the gravitational pull that would otherwise slow it down. Despite its significance, scientists still know very little about dark energy and its mechanisms. However, it is believed to constitute most of the universe total energy.

6. Cosmic waves are energy signals propagating through space, produced by violent events in the universe, such as black hole collisions or star mergers. As they travel through space-time, these waves carry valuable information about the celestial events that created them.

REFERENCES

Alharbi, W. S. & Rashid, M. [2022], *A Review of Deep Learning Applications in Human Genomics Using Next-Generation Sequencing Data*, Human Genomics, 16, n. 1, pp. 1–20.

Caligiore, D., Giocondo, F. & Silvetti, M. [2022], *The Neurodegenerative Elderly Syndrome (NES) Hypothesis: Alzheimer and Parkinson are Two Faces of the Same Disease*, IBRO Neuroscience Reports, 13, pp. 330–343.

Chell, E. & Athayde, R. [2011], *Planning for Uncertainty: Soft Skills, Hard Skills and Innovation*, Reflective Practice, 12, n. 5, pp. 615–628.

Epel, E. S., Blackburn, E. H., Lin, J., Dhabhar, F. S., Adler, N. E., Morrow, J. D. & e Cawthon, R. M. [2004], *Accelerated Telomere Shortening in Response to Life Stress*, Proceedings of the National Academy of Sciences, 101, n. 49, pp. 17312–17315.

Heckman, J. J. & Kautz, T. [2012], *Hard Evidence on Soft Skills*, Labour Economics, 19, n. 4, pp. 451–464.

Lumera, D. & De Vivo, I. [2020], *Biologia della gentilezza*, Milano, Mondadori.

Neyrinck, M., Elul, T., Silver, M., Mallouh, E., Aragón-Calvo, M., Banducci, S., ... & Zahr, N. M. [2020], *Exploring Connections Between Cosmos & Mind through Six Interactive Art Installations* in As Above As Below, arXiv, 2008.05942.

Pfeifer, S., McCarthy, I. G., Stafford, S. G., Brown, S. T., Font, A. S., Kwan, J., ... & Schaye, J. [2020], *The Bahamas Project: Effects of Dynamical Dark Energy on Large-Scale Structure*, Monthly Notices of the Royal Astronomical Society, 498, n. 2, pp. 1576–1592.

Vazza, F. & Feletti, A. [2020], *The Quantitative Comparison between the Neuronal Network and the Cosmic Web*, Frontiers in Physics, 8, p. 491.

Vilhekar, R. S. & Rawekar, A. [2024], *Artificial Intelligence in Genetics*, Cureus, 16, n. 1, pp. 1–9.

Conclusions

The pace of the artificial intelligence (AI) revolution is accelerating across all sectors, perhaps to an excessive degree. Even for professionals in the field, keeping up with the latest algorithmic developments can be challenging. There is a risk that the primary focus of developers and manufacturers of AI-driven devices may shift towards pushing technological boundaries to launch the newest applications and beat competitors to market. Adopting this mindset can be dangerous, as it risks losing sight of the true objective of technological advancement: Enhancing human well-being. This focus is particularly critical in the medical field. Our responsibility is to ensure that AI-driven changes occur in alignment with ethical principles. An AI application must be ethically sound rather than merely technically feasible or economically profitable. In other words, just because we have the technical capability to create an AI application does not mean we should do so without conducting thorough ethical evaluations.

In these concluding pages, I will address several cross-cutting issues related to the topics discussed in previous chapters. The focus is promoting the ethical development of AI in healthcare. The first issue concerns the mimetic nature of AI. Historically, most technologies developed by humans served a *prosthetic function*, enhancing or extending human capabilities. However, recent advances in AI, particularly generative AI, have shifted toward a *mimetic function*. This means that, rather than merely augmenting human abilities, AI now mimics, imitates, or replicates human behaviors and creative processes [Campione et al. 2024]. Generative AI, for instance, can create new text, images, music, and more – producing content that closely resembles what a human might generate. Unlike previous AI models designed for specific tasks, such as speech recognition or translation, generative AI leverages vast training data to produce novel outputs that reflect a learned "understanding" of the world. This ability to *simulate human-like creativity and interaction* can be observed in AI ability to

DOI: 10.1201/9781003606130-5

respond to questions, invent stories, and even express emotions in a way that seems genuine. Nevertheless, despite its capacity to mimic human behavior, generative AI lacks the intrinsic motivations, consciousness, and intentionality that define human intelligence. In humans, motivation stems from personal needs, desires, and purposes. By contrast, generative AI is programmed to respond to user inputs but does not have intrinsic motivation. Human beings perceive themselves as unique individuals. They are also aware of how their thoughts, actions, and surroundings are interconnected. In contrast, generative AI has no self-awareness or understanding of the world. Furthermore, humans act intentionally, making decisions combining knowledge, values, goals, and desires. Human intentionality allows for purposeful actions directed toward specific outcomes. Generative AI, however, operates purely in response to specific inputs and cannot perform actions independently or with intent; it can only act based on user instructions or pre-set conditions.

Understanding the mimetic nature of AI is essential for developing an ethical and responsible approach to its application, especially in medicine. Since AI systems can replicate patterns found in their training data, they may also inherit biases from this data (see Chapter 3, Section 3.4). For example, if the data used to train diagnostic algorithms includes biased information related to race or gender, AI could perpetuate and even amplify these biases, resulting in unfair or discriminatory decisions in medical care. Moreover, as AI systems are often complex and opaque, recognizing their mimetic nature can encourage developers to focus on making AI decision-making processes more transparent and explainable (see Chapter 3, Section 3.5). Ensuring that AI-generated decisions are understandable is crucial. Awareness of AI tendency to mimic human behaviors can also help prevent misuse, such as the spread of misinformation or the creation of harmful content designed to manipulate people.

Ensuring that AI-based medicine is accessible to everyone, regardless of origin or economic status, is critical to fostering ethically sustainable technological progress and upholding the principles of equality and social justice. Equal access to medical technology promotes fairness in healthcare, ensuring that all individuals, regardless of financial means or geographic location, receive the same opportunities for advanced medical care. This aspect is especially crucial for individuals in low-income communities or developing countries, who must not be excluded from the advanced medical innovations readily accessible in wealthier nations.

Aside from social advantages, expanding access to medical AI also has technical benefits, as it directly impacts the effectiveness of these systems. AI behavior largely depends on the data used for training – essentially, its "experience". Using diverse data from different populations (across socio-economic statuses, lifestyles, etc.) enhances the AI ability to generalize and respond more accurately to various types of patients. This feature is essential since diseases and treatment responses can vary significantly across different groups. Additionally, broad accessibility can foster greater public trust and acceptance of AI in healthcare, preventing perceptions of inequity that could erode confidence in these technologies.

This book highlights the potential impact of AI on the future of medicine, but other fields, such as genetic research and regenerative medicine, will also play transformative roles. Genetic modification through genetically modified organisms (GMOs) is poised to have a significant medical impact. Genetic engineering techniques modify the genetic material of GMOs, enabling them to develop traits or capabilities they would not naturally possess. These modifications have advanced our understanding of genes and their functions, paving the way for breakthroughs such as personalized gene therapies (see Chapter 4, Section 4.2). *Gene editing*, which can correct or replace defective genes, is especially promising for treating inherited genetic disorders. For example, diseases caused by mutations in a single gene, like cystic fibrosis, sickle cell anemia, or color blindness, could be cured by correcting the faulty gene. In sickle cell anemia, where oxygen transport is impaired, gene editing can deal with the disorder by replacing the DNA sequence that causes the condition. However, the real challenge lies in applying these techniques to more common and complex diseases, which often involve multiple genes [Barbujani 2019]. Scientists can engineer GMOs to produce complex drugs more efficiently and cheaply than traditional methods, making therapies more accessible. These organisms allow the rapid production of large quantities of proteins or enzymes, which benefits drugs that require mass production. Unlike conventional approaches, which often rely on extracting substances from plants or animals, GMOs synthesize these compounds without depleting natural resources, reducing environmental impact. GMOs could reduce the risk of disease transmission associated with drugs produced from animal cells or tissues. The controlled environment in which GMOs create these substances ensures safety. GMOs also enable the cultivation of human tissues and the creation of compatible organs, helping reduce transplant waiting times. Of course, GMO research also raises ethical concerns that must be addressed to ensure its medical applications are safe, effective, and ethically sound.

Regenerative medicine offers a transformative approach to healing by focusing on repairing, replacing, or renewing damaged tissues using techniques such as cell therapies, tissue engineering, and molecular medicine [Cossu 2018]. This approach could replace traditional drugs with modified viruses that carry therapeutic genes instead of pathogens. Additionally, regenerative medicine holds promise for repairing tissues damaged by injury or disease, such as replacing nerve cells in spinal cord injuries. It also has the potential to produce lab-grown organs from a patient's own cells, reducing reliance on donors and minimizing the risk of immune rejection. Furthermore, research into regenerative medicine may provide insights into aging, offering the possibility of slowing or even reversing age-related changes, which could improve longevity and quality of life. However, despite its vast potential, there are still significant ethical, scientific, and logistical hurdles to overcome before regenerative medicine can be widely implemented and accessible to all.

The synergy between GMOs, regenerative medicine, and AI presents an exciting and largely unexplored frontier in medical innovation. Together, these fields promise to reshape disease treatment, enhance our understanding of biological systems, and improve overall health outcomes. This intersection is poised to become the core of the next transformative wave in medicine.

REFERENCES

Barbujani, G. [2019], *Sillabario di genetica per principianti*, Milano, Bompiani.

Campione, F., Catena, E., Schirripa, A. & Caligiore, D. [2024], *Creatività Umana E Intelligenza Artificiale Generativa: Similarità, Differenze E Prospettive*, Sistemi Intelligenti, n. 1, pp. 1–26.

Cossu, G. [2018], *La trama della vita. La scienza della longevità e la cura dell'incurabile tra ricerca e false promesse*, Venezia, Marsilio.

Index

Note: **Bold** page numbers refer to **boxes** and *Italic* page numbers refer to *figures*.

A

Alzheimer's disease, 20, 21, 33–37, 45, **56**, 59, 69, 72, 77, *78–80*, 86, 93

C

ChatGPT, **52–54, 55, 56**
Classification, 7, 19, **71**, 81
Clustering, 7, 19
Computational phenotyping, 26

D

Data-driven approach, 19–*20*, 21, 23, 24, 27, 40, 76
Digital twin, xi, 21, 23–24, *25–29*, 32, 33, 38, 80, 82, 83, 87
DNA, 38, 43, 75, 76, 81, 82, 93, 97
Dopamine, 22, 27, 34, 37, *39–41*, *78*, 79

E

EHR (Electronic Health Records), 9, 10, 66
Explainability, 5, 6, 68–**71**, 72

F

Features, 4–6, 16, 17, 21, 33
  extraction, 4, 5
  importance, 21, 24, 33, **71**, 81

G

GDPR (General Data Protection Regulation), 66
Genetic, x, xi, 10, 35, 38, 66, 69, 75, 76, 77, *78*, 79, 81, 82, 87, 92, 93, 97
  data, xi, 66, 69, 87
  factors, x, xi, 75, 76, *78*
  information, 35, 38, 92
  markers, 70
  mutations, 38, 45, 81, 82
  profile, 38, 43, 69
  sequences, 69
  variability, x, 10
  variants, 69, 70, 81
GMOs (genetically modified organisms), 97

H

HIPAA (Health Insurance Portability and Accountability Act), 66

L

LLM (Large Language Model), 54

M

Metaverse, 28–32, 62, 67

N

NES (Neurodegenerative Elderly
　　Syndrome), 77, *78*, 79
Neuralink, 59, 60, 73
Neurorights, 59

O

Organoids, 32, 33

P

Parkinson's disease, 22, 23, 25, 26, 34, 39,
　　41, **56**, 59, 63, 70, 77, *78*, 79, 86, 93
Personalized medicine, x, xi, 10, 15
Predictive coding, 30

R

Regenerative medicine, 97, 98
Regression, 7, 17, **71**, 83, 91, 92
　　linear, 17, **71**
　　logistic, **71**
Reinforcement learning, **8**, 24

S

Serotonin, 22, 39, 40, *41*, 77, *78*, 79
Subsymbolic approach, 2–6, 9
Supervised learning, **7**, 24
Symbolic approach, 2–6, 9
Systemic, 23
　　disease, 34, 36
　　mechanisms, 87
　　perspective, 23
　　therapies, 38

T

Telomeres, 75, 76
Theory-driven approach, 19, 21–24, 26, 27,
　　37, 39

U

Unsupervised learning, **7–8**, 24, 38